ENERGY POLICIES, POLITICS AND PRICES

INTERNATIONAL REVIEW OF ENERGY EFFICIENCY STANDARDS AND LABELING PROGRAMS

ENERGY POLICIES, POLITICS AND PRICES

Additional books in this series can be found on Nova's website under the Series tab.

Additional E-books in this series can be found on Nova's website under the E-book tab.

ENERGY POLICIES, POLITICS AND PRICES

INTERNATIONAL REVIEW OF ENERGY EFFICIENCY STANDARDS AND LABELING PROGRAMS

CHAD R. SCHMITZ
EDITOR

New York

For permission to use material from this book please contact us:
Telephone 631-231-7269; Fax 631-231-8175
Web Site: http://www.novapublishers.com

Library of Congress Cataloging-in-Publication Data

ISBN: 978-1-62948-002-2

Published by Nova Science Publishers, Inc. † New York

CONTENTS

PREFACE

Appliance energy efficiency standards and labeling (S&L) programs have been important policy tools for regulating the efficiency of energy-using products for over 40 years and continue to expand in terms of geographic and product coverage. The most common S&L programs include mandatory minimum energy performance standards (MEPS) that seek to push the market for efficient products, and energy information and endorsement labels that seek to pull the market. This book seeks to review and compare some of the earliest and most well-developed S&L programs in three countries and one region: the U.S. MEPS and ENERGY STAR, Australia MEPS and Energy Label, European Union MEPS and Ecodesign requirements and Energy Label and Japanese Top Runner programs. For each program, key elements of S&L programs are evaluated and comparative analyses across the programs undertaken to identify best practice examples of individual elements as well as cross-cutting factors for success and lessons learned in international S&L program development and implementation.

Chapter 1 - Appliance energy efficiency standards and labeling (S&L) programs have been important policy tools for regulating the efficiency of energy-using products for over 40 years and continue to expand in terms of geographic and product coverage. The most common S&L programs include mandatory minimum energy performance standards (MEPS) that seek to push the market for efficient products, and energy information and endorsement labels that seek to pull the market. This study seeks to review and compare some of the earliest and most well-developed S&L programs in three countries and one region: the U.S. MEPS and ENERGY STAR, Australia MEPS and Energy Label, European Union MEPS and Ecodesign requirements and Energy Label and Japanese Top Runner programs. For each program, key elements of S&L programs are evaluated and comparative analyses across the programs undertaken to identify best practice examples of individual elements as well as cross-cutting factors for success and lessons learned in international S&L program development and implementation.

The international review and comparative analysis identified several overarching themes and highlighted some common factors behind successful program elements. First, standard-setting and programmatic implementation can benefit significantly from a legal framework that stipulates a specific timeline or schedule for standard-setting and revision, product coverage and legal sanctions for noncompliance. Second, the different MEPS programs revealed similarities in targeting efficiency gains that are technically feasible and economically justified as the principle for choosing a standard level, in many cases at a level that no product on the current market could reach. Third, detailed survey data such as the U.S. Residential Energy Consumption Survey (RECS) and rigorous analyses provide a strong foundation for standard-setting while incorporating the participation of different groups of stakeholders further strengthen the process. Fourth, sufficient program resources for program implementation and evaluation are critical to the effectiveness of standards and labeling programs and cost-sharing between national and local governments can help ensure adequate resources and uniform implementation. Lastly, check-testing and punitive measures are important forms of enforcement while the cancellation of registration or product sales-based fines have also proven effective in reducing non-compliance.

The international comparative analysis also revealed the differing degree to which the level of government decentralization has influenced S&L programs and while no single country has best practices in all elements of standards and labeling development and implementation, national examples of best practices for specific elements do exist. For example, the U.S. has exemplified the use of rigorous analyses for standard-setting and robust data source with the RECS database while Japan's Top Runner standard-setting principle has motivated manufacturers to exceed targets. In terms of standards implementation and enforcement, Australia has demonstrated success with enforcement given its long history of check-testing and enforcement initiatives while mandatory information-sharing between EU jurisdictions on compliance results is another important enforcement mechanism. These examples show that it is important to evaluate not only the drivers of different paths of standards and labeling development, but also the country-specific context for best practice examples in order to understand how and why certain elements of specific S&L programs have been effective.

Chapter 2 - Over the last twenty years, with growing policy emphasis on improving energy efficiency and reducing environmental pollution and carbon emissions, China has implemented a series of new minimum energy performance standards (MEPS) and mandatory and voluntary energy labels to improve appliance energy efficiency. As China begins planning for the next phase of standards and labeling (S&L) program development under the 12th Five Year Plan, an evaluation of recent program developments and future directions is needed to identify gaps that still exist when compared with international best practices. The review of China's S&L program development and implementation in comparison with major findings from international experiences reveal that there are still areas of improvement, particularly when compared to success factors observed across leading international S&L program.

China currently lacks a formalized regulatory process for standard-setting and do not have any legal or regulatory guidance on elements of S&L development such as stakeholder participation or the issue of legal precedence between conflicting national, industrial and local standards. Consequently, China's laws regarding standard-setting and management of the mandatory energy label program could be updated, as they have not been amended or revised recently and no longer reflects the current situation.

While China uses similar principles for choosing target products as the U.S., Australia, EU and Japan, including high energy-consumption, mature industry and testing procedure and stakeholder support, recent MEPS revisions have generally aimed at only eliminating the bottom 20% efficiency of the market. Setting a firm principle based on maximizing energy savings that are technically feasible and economically justified may help improve the stringency of China's MEPS program and reduce the need for frequent revisions.

China also lacks robust survey data and relies primarily on market research data in relatively simple techno-economic analyses used to determine its efficiency standards levels rather than the specific sets of analyses and tools used internationally. Based on international experiences, inclusion of more detailed energy consumption surveys in the Chinese national census surveys and statistical reporting systems could help provide the necessary data for more comprehensive standard-setting analyses.

In terms of stakeholder participation in the standards development process, stakeholder participation in China is limited to membership on technical committees responsible for developing or revising standards and generally do not include environmental groups, consumer associations, utilities and other NGOs. Increasing stakeholder involvement to broader interest groups could help garner more support and feedback in the S&L implementation process.

China has emerged as a leader in a national verification testing scheme with complementary pilot check-testing projects, but it still faces challenges with insufficient funding, low local awareness amongst some regulatory agencies and resistance to check-testing by some manufacturers, limited product sampling scope, and testing inconsistency and incomparability of results. Thus, further financial and staff resources and capacity building will be needed to overcome these remaining challenges and to expand impacts evaluations to assess the actual effectiveness of implementation and enforcement.

In: International Review ...
Editor: Chad R. Schmitz
ISBN: 978-1-62948-002-2

Chapter 1

INTERNATIONAL REVIEW OF THE DEVELOPMENT AND IMPLEMENTATION OF ENERGY EFFICIENCY STANDARDS AND LABELING PROGRAMS*

Nan Zhou, Nina Zheng and David Fridley

EXECUTIVE SUMMARY

Appliance energy efficiency standards and labeling (S&L) programs have been important policy tools for regulating the efficiency of energy-using products for over 40 years and continue to expand in terms of geographic and product coverage. The most common S&L programs include mandatory minimum energy performance standards (MEPS) that seek to push the market for efficient products, and energy information and endorsement labels that seek to pull the market. This study seeks to review and compare some of the earliest and most well-developed S&L programs in three countries and one region: the U.S. MEPS and ENERGY STAR, Australia MEPS and Energy Label, European Union MEPS and Ecodesign requirements and Energy Label and Japanese Top Runner programs. For each program, key elements of S&L programs are evaluated and comparative analyses across the programs undertaken to identify best practice examples of individual elements as well as cross-

* This is an edited, reformatted and augmented version of Lawrence Berkeley National Laboratory, Publication No. LBNL-5407E, dated February 2012.

cutting factors for success and lessons learned in international S&L program development and implementation.

The international review and comparative analysis identified several overarching themes and highlighted some common factors behind successful program elements. First, standard-setting and programmatic implementation can benefit significantly from a legal framework that stipulates a specific timeline or schedule for standard-setting and revision, product coverage and legal sanctions for noncompliance. Second, the different MEPS programs revealed similarities in targeting efficiency gains that are technically feasible and economically justified as the principle for choosing a standard level, in many cases at a level that no product on the current market could reach. Third, detailed survey data such as the U.S. Residential Energy Consumption Survey (RECS) and rigorous analyses provide a strong foundation for standard-setting while incorporating the participation of different groups of stakeholders further strengthen the process. Fourth, sufficient program resources for program implementation and evaluation are critical to the effectiveness of standards and labeling programs and cost-sharing between national and local governments can help ensure adequate resources and uniform implementation. Lastly, check-testing and punitive measures are important forms of enforcement while the cancellation of registration or product sales-based fines have also proven effective in reducing non-compliance.

The international comparative analysis also revealed the differing degree to which the level of government decentralization has influenced S&L programs and while no single country has best practices in all elements of standards and labeling development and implementation, national examples of best practices for specific elements do exist. For example, the U.S. has exemplified the use of rigorous analyses for standard-setting and robust data source with the RECS database while Japan's Top Runner standard-setting principle has motivated manufacturers to exceed targets. In terms of standards implementation and enforcement, Australia has demonstrated success with enforcement given its long history of check-testing and enforcement initiatives while mandatory information-sharing between EU jurisdictions on compliance results is another important enforcement mechanism. These examples show that it is important to evaluate not only the drivers of different paths of standards and labeling development, but also the country-specific context for best practice examples in order to understand how and why certain elements of specific S&L programs have been effective.

1. Introduction

Appliance energy efficiency standards and labeling (S&L) programs have been important energy efficiency policy tools for regulating the efficiency of energy-using products for over 40 years and have continued to expand in terms of geographic and product coverage. As a mandatory policy, minimum energy performance standards (MEPS) help push the efficiency of products on the market by setting energy efficiency metrics that must be met and help eliminate inefficient products that cannot meet the standard. At the same time, mandatory and/or voluntary energy information and endorsement labels seek to pull the market for efficient products by providing information for consumers to identify and/or compare the energy efficiency of similar product models in their purchase decision-making. While S&L programs have been in effect for over three decades in most developed countries and regions, they have only recently begun receiving policy attention in rapidly growing and developing countries.

This study seeks to review and compare some of the earliest and most well-developed S&L programs in three countries (the United States, Australia and Japan) and one region (the European Union) that have often been considered leaders in S&L program development and implementation. Specifically, this study provides in-depth review and comparative analysis of the development and recent advancements in the U.S. MEPS and ENERGY STAR program, Australia MEPS and Energy Label, European Union MEPS, Ecodesign requirements and Energy Label, and Japanese Top Runner program. For each national program, the review and comparative analysis focus on key elements of S&L development including legal framework for related policies, standard-setting processes, analyses and data needs, stakeholder participation, program implementation and enforcement mechanisms, basis for test procedures and overall program resources. The in-depth programmatic review and comparative analysis of each program provides the foundation for identifying best practice examples of each programmatic element as well as cross-cutting factors for success and lessons learned in S&L program development and implementation.

2. Legal Framework

2.1. United States

The United States (U.S.) minimum energy efficiency standards program was officially established in 1975 by Part B of Title III of the Energy Policy and Conservation Act (EPCA) of 1975. The 1975 EPCA set test procedures, conservation targets and lays out the process for prescribing efficiency standards if targets are not set. EPCA also stated that federal regulations will supersede any state regulation on equipment energy efficiency. It directs the Federal Trade Commission to prescribe energy consumption labeling rules for covered products, thereby leading to the creation of the mandatory EnergyGuide information label. In 1978, the National Energy Policy Conservation Act (NEPCA) directed the Secretary of Energy to establish mandatory MEPS that meets maximum improvement in energy efficiency which is technologically feasible and economically justified for specified appliances and classes of industrial equipment. It also provided a statutory foundation for states to petition the Secretary of Energy for an exemption from being superseded by federal regulation given certain conditions. NEPCA was followed by the National Appliance Energy Conservation Act (NAECA) of 1987, which set forth specific standards for covered products and set deadlines for the Department of Energy to issue the standards. In addition, the ENERGY STAR voluntary labeling program was introduced as a joint program of the U.S. Environmental Protection Agency (EPA) and Department of Energy (DOE) in 1992. The product coverage of the U.S. MEPS program was further expanded to include certain commercial and industrial equipment by the Energy Policy Act of 1992 and through 15 new standards and 11 test procedures adopted in the 2005 Energy Policy Act. The 2005 Energy Policy Act also authorized budget appropriations of USD $90 million per fiscal year for 2006 through 2010 for standards development and all the other provisions (e.g., industrial, transportation, renewable fuel standard, etc.). Most recently, the Energy Independence and Security Act (EISA) of 2007 added 18 new energy conservation standards, including standards for 10 new products, 16 new or revised test procedures and mandated regular rulemaking reviews every six years for MEPS and seven years for test procedures. Figure 1 illustrates the legislative timeline related to the U.S. Department of Energy's standards program.

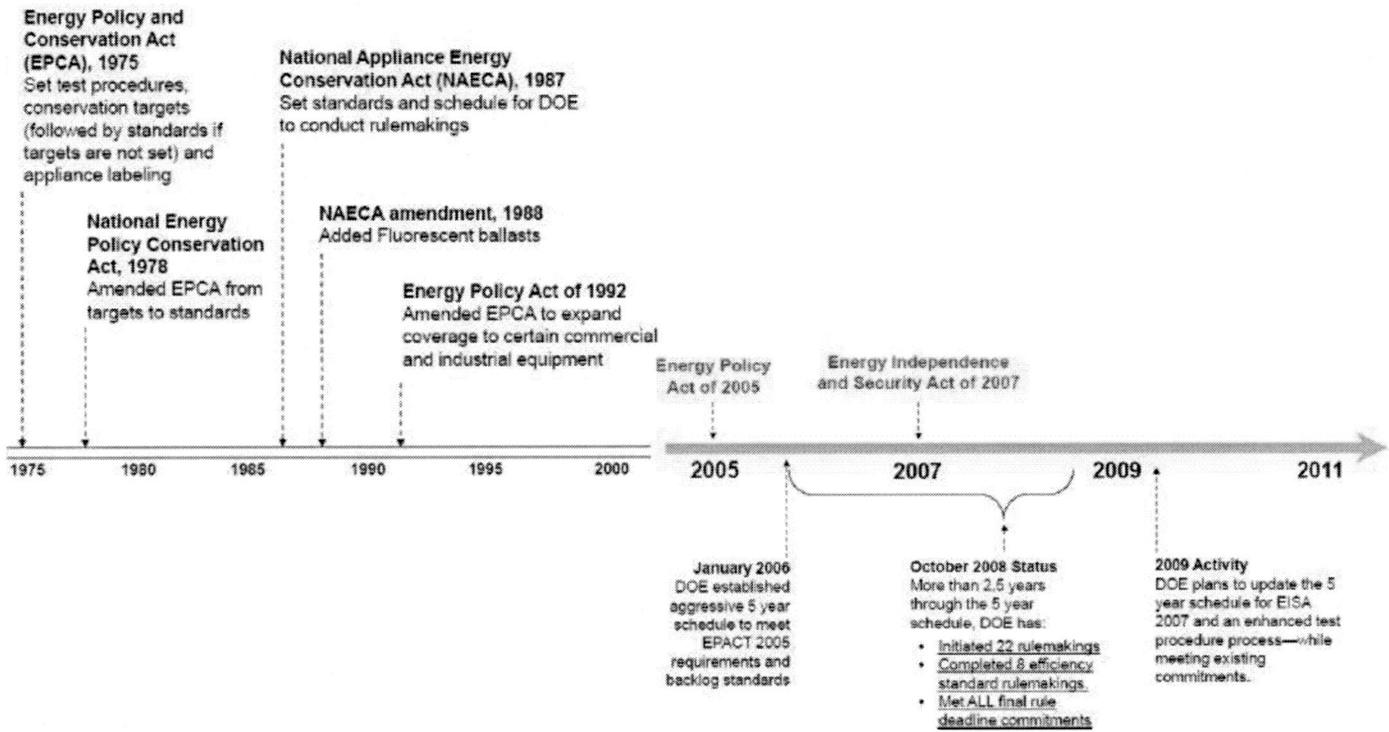

Sources: Rodgers, D. E., 2008, "US DOE Appliance Standards Program Challenges and Opportunities for Global Coordination." Presented at ECEEE Production Efficiency 2008 Conference. Brussels: October 30-31, 2008. Waide, P., 2010, "Opportunities for Success and CO2 Savings from Appliance Energy Efficiency Harmonization." CLASP Report (Draft).

Figure 1. Legislative Timeline of U.S. Standards Program, 1975 to present.

2.2. Australia

The Australian energy labeling program was created out of 1979 proposals by the two states of New South Wales (NSW) and Victoria to establish energy labeling for refrigerators and freezers and national discussions throughout the early 1980s. After failed negotiations on creating a voluntary labeling program with manufacturers and trade associations, NSW and Victoria unilaterally established mandatory energy labels within their jurisdiction in 1986. The South Australian state government also established a mandatory label in 1990, thereby extending the population coverage of the three labeling programs to two-thirds of the national total. In recognizing the disadvantages of a patchwork of labeling programs, the National Appliance and Equipment Energy Efficiency Committee (NAEEEC) was created in 1992 to facilitate the introduction of a national MEPS program. MEPS implementation was further driven by the 1998 National Greenhouse Strategy, which explicitly mandated that "improvements in the energy efficiency of domestic appliances and commercial and industrial equipment will be promoted by extending and enhancing the effectiveness of existing energy labeling and minimum energy performance standards programs" in section 4.10 (Commonwealth of Australia, 1998).

By 1999, all jurisdictions had implemented State and Territory regulations to enact mandatory energy labeling and MEPS program and a restructured approach for MEPS was introduced in the first national administrative guidelines for MEPS and labeling programs in 2000. After 2000, state and territorial standards and labeling activities were managed and coordinated by the NAEEEC, which later became the Equipment Energy Efficiency (E3) Committee. As part of the coordination activities and to provide a basis for ensuring consistency across jurisdictions, state and territory legislation were asked to reference Australian Standards for the technical content related to MEPS and labeling. Specifically, Australia Standards were made uniform with part 1 of the standard detailing the specific test procedures and part 2 detailing the technical requirements for MEPS and labeling. Nevertheless, in the case of conflict between the national guideline and jurisdictional legislation, the state or territory legislation supersedes the national guideline in that local standards can be more stringent than national standards.. As of 2010, Australia has 17 MEPS, 7 mandatory comparative information labels and 2 voluntary labels covering 18 product groups with over 17,000 registered models (E3 2011). Forward plans are published every three years by the E3 Committee to set forth a calendar for introducing new and revised MEPS and labeling requirements, in addition to Ten Year Strategies that are produced for several key areas of end-uses to outline products and measures that the government intends to pursue (Ellis 2012).

2.3. European Union

The legal framework for the European Union (EU)'s MEPS and labeling programs is built on the foundation of two key European Commission directives in the early to mid-1990s and has expanded more recently to include new Ecodesign requirements (Figure 2). First, Directive 1992/75/EEC introduced mandatory comparative energy information labeling for household refrigerators, washing machines and dryers, dishwashers, ovens, water heaters and hot water storage, lighting and air conditioners. The EU energy label ranked a product's annual energy consumption relative to other similar models from grade A (the most efficient) to grade F (the least efficient). Under this directive, suppliers are responsible for providing accurate technical documentation for the label's information. Member states then have to ensure suppliers fulfill their labeling obligations, prohibit other labels that do not comply with the labeling requirements or may mislead or confuse consumers and initiate educational and promotional campaigns to support the labels. Second, the 1996 Directive 96/57EC for energy efficiency requirements of household electric refrigerators, freezers and combinations required member states to adopt and publish laws, regulations and administrative provisions to introduce and enforce national MEPS programs that meet the directive requirements within one year.

More recently, Directive 2005/32/EC introduced the framework for setting new Ecodesign requirements for Energy-using Products complementary to the EU Energy Label and integrated with existing MEPS directives. Specifically, the Ecodesign directive sets up a framework for determining specific standards or targets for lifetime performance criteria with new emphasis on the energy consumption and environmental aspects of the non-use phases of energy-using products in the domestic, commercial and industrial sectors. This directive was followed in 2008 by the Ecodesign Framework Directive which sets up the 2009-11 implementation plan for the Ecodesign Directive and states binding requirements shall be set by implementing measures specific to a product group in the absence of any valid self-regulatory initiatives. The 2008 framework directive also sets out an indicative list of product groups to be considered as priorities for adopting implementing measures.

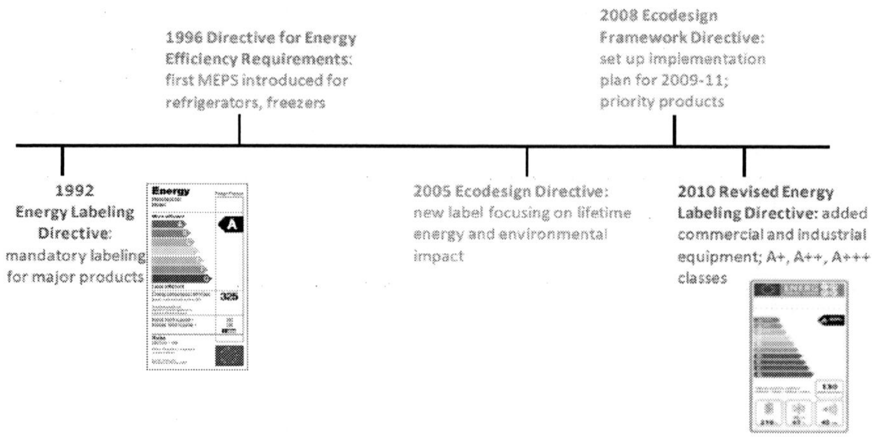

Figure 2. Timeline of Major EU Directives on MEPS and Labeling.

2.4. Japan

Following the 1970s oil crises, Japan adopted its Energy Conservation Law (ECL) in June 1979 which included guidelines for regulating the energy consumption of high energy consuming equipment. The ECL stipulated that energy standards will apply to both importers and domestic manufacturers and mandatory labeling must include product name, model, energy efficiency and consumption of each designated product. Japan's ECL was amended six times between 1979 and 2008 to reflect growing concern with energy efficiency improvement and the enactment of the Kyoto Protocol, with two of the six amendments providing specific provisions for standards and labeling. For instance, the 1998 amendment gave the Ministry of Economy, Trade and Industry (METI) the authority to set standards for energy consumption of designated machinery and equipment that are "heavily used in Japan and consumes a considerable amount of energy", including refrigerators, air conditioners and passenger cars (Government of Japan, 1998). The Top Runner standards program was introduced in Section 6 of the amendment, which stipulates that manufacturers are obligated to make efforts to improve the energy efficiency of their equipment. In the 2008 Revision of the Energy Conservation Law, Article 78 specified that the standards should be set based on the highest level of performance while taking into consideration future prospects for technological development and should be revised as necessary. The ECL also gave METI the authority to use public announcements and orders against noncompliant manufacturer as an enforcement measure. In order to comply with the ECL labeling requirements, Japan's voluntary Energy Saving Labeling Program was launched on August 21, 2000 and covered 16 products by 2008 (Figure 3). For each product model, the label provides consumers with information on the applicable Top Runner target year, the particular model's achievement rate relative to the Top Runner target, and its annual energy consumption. The model also features a green "e" mark for products that achieve over 100% of the target and an orange "e" mark for products that do not achieve the target. After the Kyoto Protocol went into force in late 2004, the 2005 ECL amendment further emphasized the importance of retailer participation in displaying information on products' energy efficiency performance to help raise consumer awareness. The 2006 Revised Law Concerning the Rational Use of Energy introduced uniform guidelines for labeling and created Japan's Uniform Energy-saving Label, which encompasses the voluntary energy label information and also provides a 5-star rating system for a product's efficiency and its estimated electricity bill. To date, the Uniform Energy Saving Label covers 5 products while the voluntary energy saving label covers 13 products with 3 more scheduled to be added.

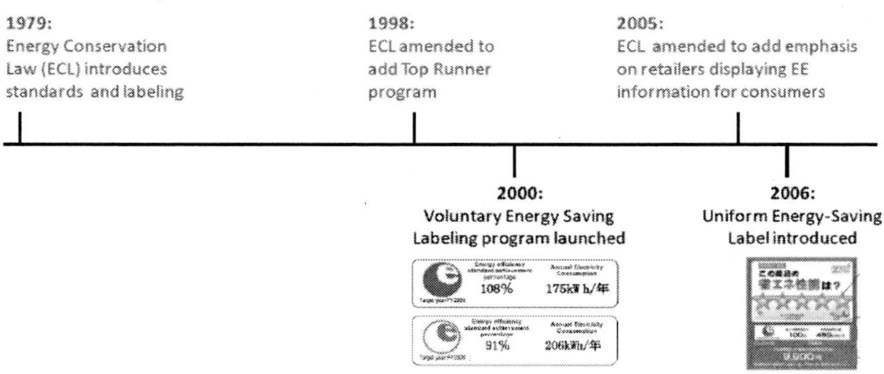

Figure 3. Timeline of Japanese Standards and Labeling Policy and Programs.

2.5. Comparison

The history of key laws and regulations that laid the foundation for MEPS and labeling programs in the U.S., Australia, EU and Japan reflect two historical phases of national and regional emphasis on energy conservation and efficiency programs. Following the 1970s oil shocks and growing concerns over dependence on imported energy sources, both the U.S. and Japan passed energy conservation laws while regional proposals for energy labels emerged in Australia. EU stands out as an exception since it was not created as a geo-political entity until 1993. Within the last two decades, renewed focus on MEPS and energy labeling as tools for slowing down energy consumption growth has emerged alongside growing recognition and concerns over climate change. Detailed policies and accelerated implementation of the standards and labeling programs thus emerged in the late 1990s and early 2000s with the introduction of EU MEPS and Energy Label, creation of Japan's Top Runner program and the mandated timetables for test procedures and MEPS in the 2005 U.S. Energy Policy Act. With the exception of the EU, MEPS or target standards were introduced and used as the basis for subsequent creation of an energy information labeling program in the three other selected countries.

Another difference in the legal framework for MEPS and labeling programs in the selected countries is whether national or regional legislations have precedence, which in turn determines the pace of standard revisions. For the U.S. and EU, granting exemptions for more stringent state MEPS regulation to take precedence over national regulations is discouraged due to concerns over trade barriers. The EU specifically prohibits member states from introducing unilateral mandatory requirements for traded goods because of trade barriers while the U.S. Secretary of Energy can only grant states exemption if state regulation is proven to not burden interstate commerce. Australia differs in that state or territory legislation set the MEPS level and that some states or territories may enact legislation with more stringent MEPS than other states. This has enabled local jurisdictions to unilaterally adopt more stringent standards with Queensland adopting higher MEPS for some classes of air conditioners in 2009 and South Australia in 2010.

Although all legislation prohibits the sale of products that do not meet MEPS criteria or labeling requirements, there is a greater range in the authorized enforcement mechanisms and legal ramifications for non-compliance in national legislation. Of the four regions, only the U.S. and Japan prescribes sanctions and measures for addressing non-compliance in its legislation which differs in the severity and form of the sanction. In the U.S., EPCA allows any person to initiate civil action against manufacturer or private labeler in violation of MEPS while NEPCA sets forth procedures for determining civil penalties for MEPS violations. Japan differs in that its 1998 amendment calls for the use of public announcements and order against non-compliant manufacturers. The EU directives assign enforcement authority to the member states and do not set forth any specific legal repercussions for non-compliance. However, the United Kingdom (UK) is currently reviewing legislation to impose civil sanctions against manufacturers that fail to meet the Ecodesign requirements. Australia does not specify any punitive measures for MEPS or labeling violations in its national administrative guidelines, but sanctions are outlined in state or territorial legislations. Sanctions may include de-registration of noncompliant products, which results in these products being prohibited from being imported or sold in the Australia market. For cases of misleading conduct in labeling (i.e., fraudulent labeling), the products may be referred to the Australia Competition and Consumer Commission and penalties to compensate for lost energy savings to consumers and environmental damages may be issued (Ellis 2012).

3. Standard Setting and Revision Process

The guiding principles and specific steps and analyses undertaken as part of the energy efficiency standard setting and revision process are crucial in influencing the subsequent impacts of the standards on energy savings and emission reductions. This can be seen in the example of top-mount auto-defrost refrigerator standards in the 1990s with a simplified[1] comparison of the approaches for actual and theoretical standards set by different countries and as compared to products on the market in. In this example, the U.S. DOE standard setting principle and process resulted in the 1993 standard being set at a much more stringent level of efficiency in 1990 with almost no products available on the market at the time of the standard-setting. Similarly, the 2001 standard was set at the level where no product on the 1993 market could reach. Compared to the U.S., the Japanese Top Runner standard and the EU standards would have been slightly less stringent but would still set the future standard at relatively high efficiency levels of existing products on the market. The guiding principles and analytical tools and methods for determining the efficiency standard thus have an important impact on the market distribution of efficient products. Experience demonstrates that industries often underestimate their potential for higher efficiency, and a technically feasible and economically viable standard can help bring the market to a much higher efficiency level.

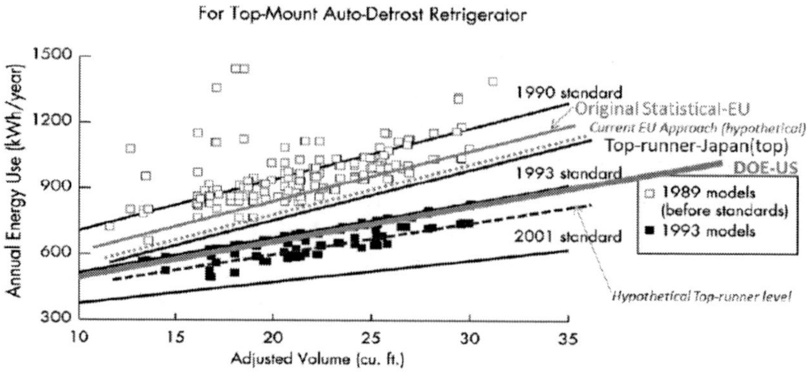

Source: Based on figure provided by Rosenquist 2010.

Note: standard levels shown in the figure are for illustrative purposes and do not necessarily reflect differences that result from different test procedures.

Figure 4. Comparison of Impact of Different Standard-setting Criteria.

3.1. United States

The U.S. guiding principle for setting the threshold for a minimum energy performance standard is to achieve maximum efficiency that is technologically feasible and economically justified, thus maximizing energy savings. The Secretary of Energy has discretion in weighing the benefits and burdens of selecting the final stringency level of the standard for a given product class, where product class is defined by differences in a given product's utility functions to consumers. In doing so, the Secretary of Energy must consider seven statutory criteria, including:

1. The economic impact of standard on consumers and manufacturers
2. Lifetime operating cost savings resulting from the standard
3. Total projected energy savings resulting from the standard
4. Impact of the standard on utility or performance of products
5. Impact of any lessening of competition likely to result from the standard
6. Need for national energy conservation
7. Other factors the Secretary consider relevant

From a procedural perspective, the U.S. has followed a rulemaking process for setting its MEPS since DOE restructured the standard setting process in a formal Process Rule in 1996. The current rulemaking process is initiated by the authority of a planned multi-year schedule of approximate rule initiation and final action dates issued by DOE. The current schedule of appliance standards development was formulated in 2005 and includes standards for 24 products. The energy standards rulemaking process officially begins with the publication of a Notice of Determination in the Federal Register to determine if a new or revised standard is needed. This is followed by a 30 calendar day comment period for the public to provide input to DOE regarding the Notice of Determination.

If DOE determines that a rulemaking should be undertaken for the given product following public input, a Framework Document is drafted by either Lawrence Berkeley National Laboratory for residential products or by Pacific Northwest National Laboratory for commercial products. This Framework Document describes DOE's plans for conducting the supporting analyses for the rulemaking and is published with a Notice of Availability that seeks further comments or data input from the public. After a public meeting is held for further public participation, DOE and its contractors perform a variety of economic and technical analyses (see Section B.2.1 for further details). The typical duration of time from the Notice of Determination to completion of the analyses is 18 months. After the supporting analysis is completed, the results are published in an Advance Notice of Proposed Rulemaking in the Federal Register with Technical Support Document and a 30 to 45 calendar day comment period begins with a public meeting held for further comments. After this comment period, additional analyses are conducted and revisions may be made based on comments. DOE then reviews all comments and address them in the Notice of Proposed Rulemaking in the Federal Register. This period typically takes 11 months. Finally, a 60 calendar day comment period follows the published Notice of Proposed Rulemaking and final revisions can be made to the analysis if necessary. Within a six month timeframe, DOE publishes the Final Rule announcing the energy efficiency standards and their effective dates. For the ENERGY STAR program, energy efficiency requirements are set in the product specifications and typically represent the top 20% efficient products on the market. Additional guiding principles for setting the ENERGY STAR specifications include significant nationwide energy savings, provide features and performance demanded by consumers, reasonable payback period for higher incremental cost of more efficient unit, broadly available and non-proprietary technologies by more than one manufacturer and verifiable energy consumption and performance. The specific processes in the specification development cycle are shown Figure 6 below.

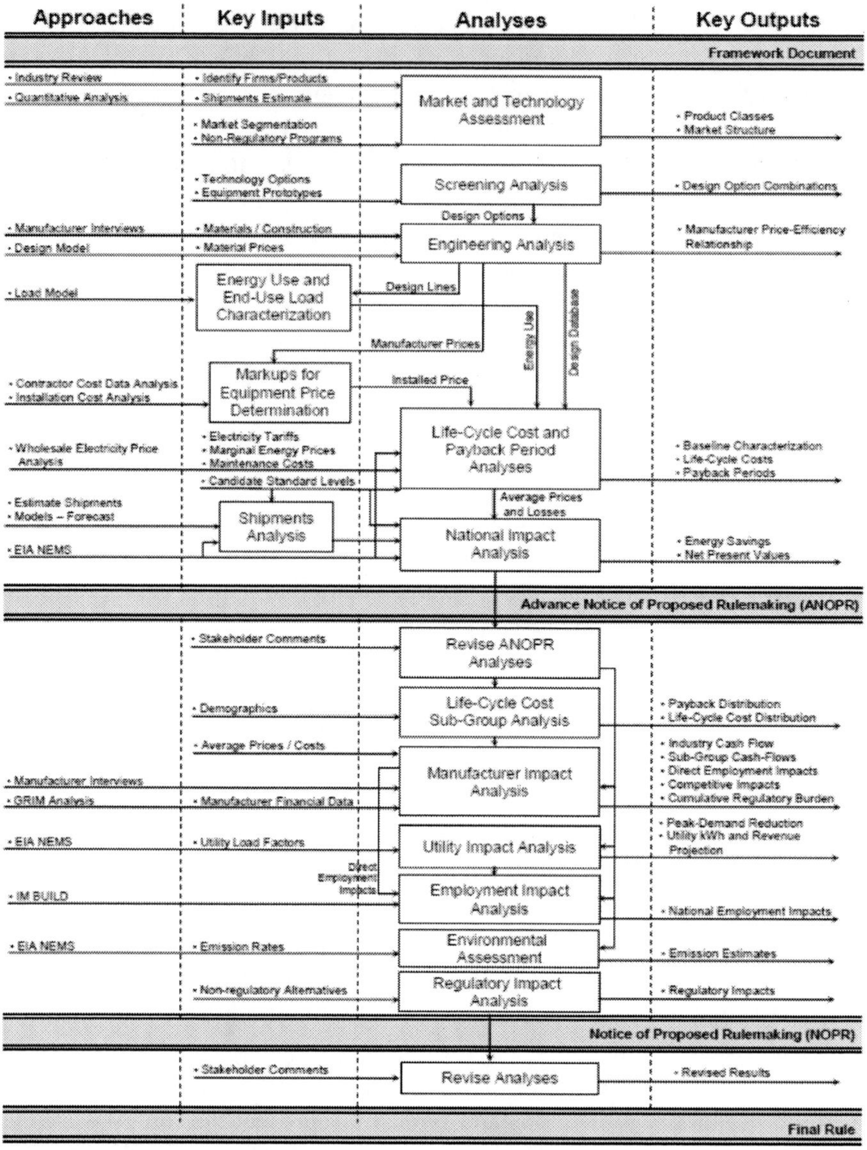

Source: U.S. Department of Energy, 2006, "Energy Conservation Standards Activities: Submitted Pursuant to Section 141 of the Energy Policy Act of 2005 and to the Conference Report (109-275) to the FY 2006 Energy and Water Development Appropriations Act." Available at: http://www1.eere.energy.gov/...standards /pdfs/implementationreport0806.pdf

Figure 5. Flow Diagram of DOE Energy Standards Rulemaking.

Source: http://www.ENERGY STAR.gov/index.cfm?c=prod development.prod development spec rev.

Figure 6. Steps in ENERGY STAR Specification Development Cycle.

Unlike the U.S. MEPS, there is no specific timeline for revising ENERGY STAR product specifications but rather, the revisions are initiated in response to changing market shares for efficient products. A general principle for considering revisions of a product specification is if the market share of ENERGY STAR qualified products in a particular category reaches 50% or higher. However, other considerations for undertaking specification revisions include changes in federal MEPS, technological changes and advancements that allow revised specifications to capture additional savings, product availability, significant issues with consumers realizing expected savings, performance or quality issues and issues with test procedures.

Most recently in May 2011, the ENERGY STAR program launched a new pilot program element to designate the most efficient or top tier efficiency models for selected product categories including clothes washers, refrigerators, televisions and heating and cooling equipment. The Most Efficient designation is intended to recognize truly exceptional, inspirational or leading edge efficiency performance and targets a very small proportion of highly efficient models such as the top 5% efficient TV models on the market.

3.2. Australia

The Australian standard setting and revision process begins with product selection in which potentially regulated products are identified by the E3 Committee. The committee is responsible for analyzing and projecting product level energy use in order to determine if identified products have significant current or projected energy use on either a per unit basis or due to high sales volume. To make this determination, the committee commissions product profiles and formulates a regulatory proposal to consider if policy intervention is necessary and if so, which policy option (MEPS, labeling, both MEPS and labeling or another policy option) is the most appropriate. The regulatory proposal is based on economic analysis, consumer research and industry research and a draft regulatory impact statement is published for public comment before the final regulatory impact statement is submitted for approval. The specific process is depicted in Figure 7.

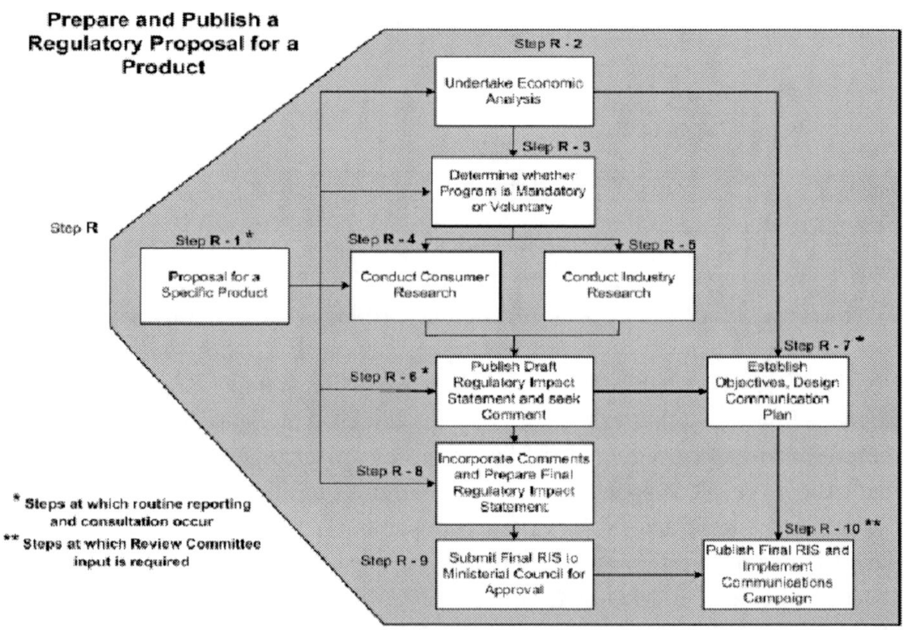

Source: E3, 2006, "The MEPS and Energy Labeling Process in Australia and New Zealand." Available at: www.energyrating.gov.au/pubs/meps-labelling-process-au-nz.pdf.

Figure 7. Processes in Preparing Regulatory Impact Statement in Australia.

Once the regulatory impact statement has been approved, further steps are taken to design and implement MEPS and/or labeling for a given product. Data collection and categorization of product classes is undertaken, followed by statistical, engineering/economic, consumer, industry, national and market analysis. As with the regulatory impact statement, a draft standard is circulated for public comment before it is finalized and published. The specific steps and processes in the formulation and implementation of an Australian MEPS are shown in Figure 8.

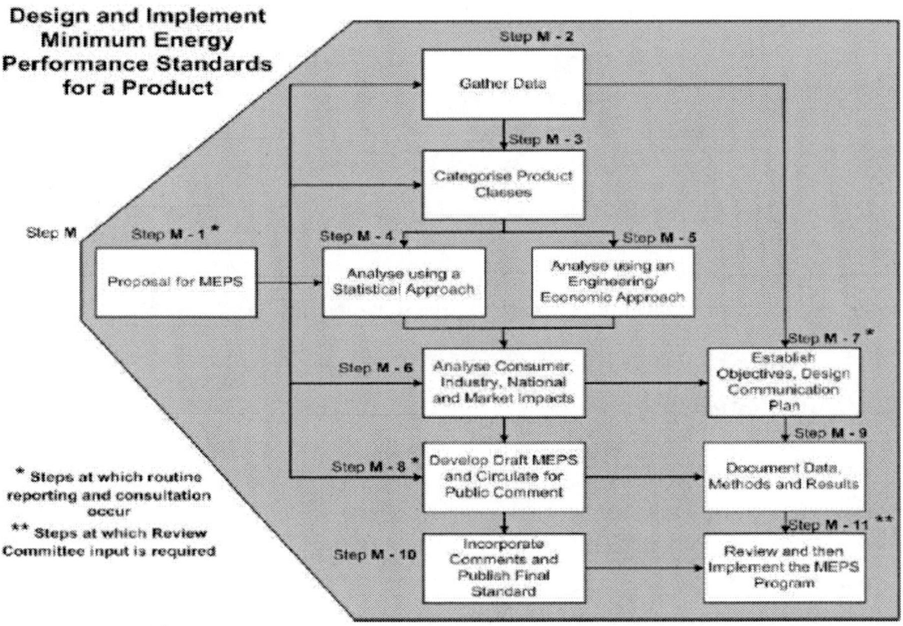

Source: E3, 2006.

Figure 8. Processes in the Design and Implementation of MEPS in Australia.

In developing energy labels, a shorter process with fewer analyses is followed (Figure 9).

Revisions of label thresholds are generally determined in several ways, including in long-term strategies formulated for the next ten years for certain products, the outcome of jurisdictional commitment to a regular review process 3 to 5 years after implementation, or identified by a reach level with or without a stated timeframe. The implementation of a revised label is usually accompanied by three specified dates: the date before which only the original label is permitted, the transition period set by state or territorial legislation in which both the original and revised labels may be used, and a compliance date after which only the revised label may be used. As previously mentioned, in rare cases where a consensus cannot be reached in negotiations between states and territories, MEPS or label revisions may also be undertaken unilaterally by local jurisdictions.

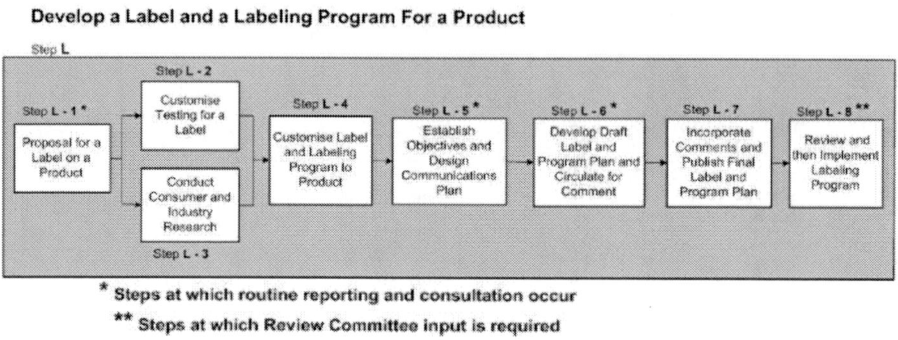

Source: E3, 2006.

Figure 9. Process for Developing Energy Label in Australia.

3.3. European Union

Similar to the U.S., EU MEPS must also be set at levels that are economically and technically justified. Directive 96/57/EC also states that the European Commission will assess results obtained from existing MEPS and decide if revisions are necessary four years after MEPS go into effect. Similarly, Directive 92/57/ECC states that an assessment of current label thresholds must be conducted three years after the application of the directive. Because the original MEPS program is now being integrated into the Ecodesign Directive, only the specific processes for the Ecodesign program will be described in detail below. The Ecodesign requirements, which cover energy efficiency but also include waste generation, consumption of other materials and resources such as water and release of hazardous substances, are currently being set on a product-by-product basis for energy-consuming and other energy-related products. Once the Ecodesign requirements for a specific product go into effect, manufacturers must meet the requirements in order to sell or trade its products within the European Economic Area.

As a new initiative with its implementation plan dictated by the 2008 Ecodesign Framework Directive, a more thorough process has been laid out for setting Ecodesign requirements that must also be economically and technically justified. More specifically, a product must first meet three basic criteria to be regulated by the Ecodesign requirements. First, the product must have significant volume and trade, measured by sales greater than 200,000 units per year within the EU Community. Second, the product must have a significant environmental impact within the Community. In the UK, this impact is defined by high primary energy consumption exceeding 1000 PJ per year with other possible indicators of water consumption, long operating time, and parts that contribute to energy consumption or expected increase in the next decade due to high growth market rate. Finally, the product must also have significant potential for improvement in environmental impact without incurring excessive costs, as indicated by market failures, the absence of policy intervention and a wide disparity in environmental performance of products with equal functionality. For the UK, this criterion can be met by energy savings potential of greater than 20% during the use phase and taking into consideration specifications in other countries and the latest information on technology development.

After a product has been determined to meet the qualifying criteria, the Methodology Study of Ecodesign of Energy-using Products known more commonly as preparatory studies are undertaken to evaluate and set the implementing measures. The specific processes and analyses included in each preparatory study are shown in Figure 10 and the specific analyses undertaken are covered in greater detail in the next section. For each product or group of products, the preparatory study begins by defining the product, existing standards and legislation following economic and market analysis. Consumer behavior analysis, local infrastructure analysis and technical analysis of existing products are also used to inform the development of a base case. The base case is then used to reflect the underlying emissions and resources in a product's life-cycle. Next, a technical analysis of best available technology serves as the basis for assessing improvement potential. Lastly, policy, impact and sensitivity analyses are conducted to evaluate the proposed implementing measure. All of the supporting documents and underlying analyses for each preparatory study process are publicly available online and can be accessed at specific websites dedicated to each preparatory study.

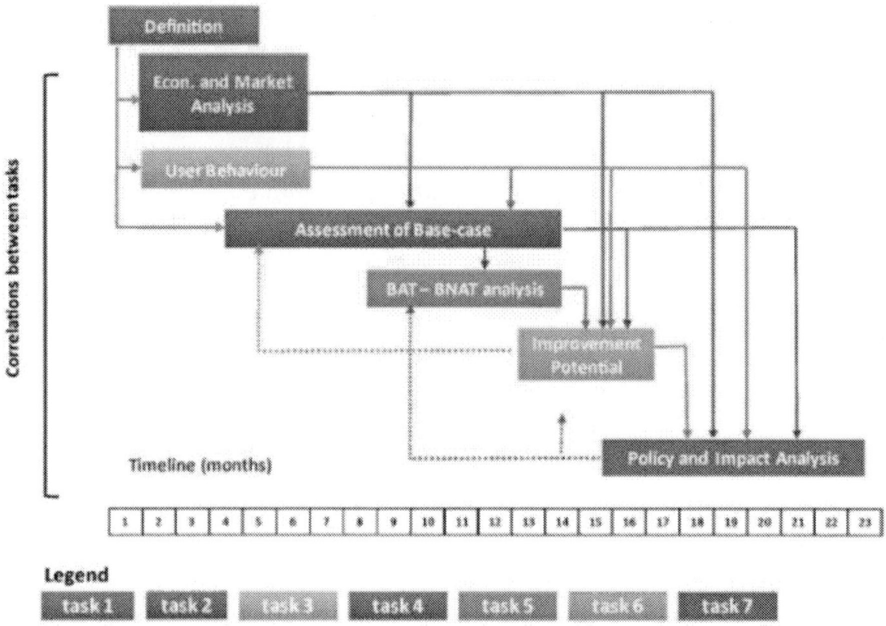

Source: Waide, P., and L. Harrington, 2010, "Opportunities for Success and CO2 Savings from Appliance Energy Efficiency Harmonization." London: CLASP Report.

Figure 10. EU Ecodesign Preparatory Study Processes.

The entire process for the preparatory study involves different research and analytical teams and occurs over a timeframe of approximately two years. Once the preparatory study has been completed, then a Consultation Forum is held for stakeholders and member states to discuss the study findings and formulate a draft regulation.

The draft regulation is submitted to the Regulatory Committee for review and upon Committee approval, is forwarded to the European Parliament for further review and the World Trade Organization is notified. Lastly, the regulation is formally adopted by the European Commission and published as a directive before entering into force. This process is illustrated in Figure 11.

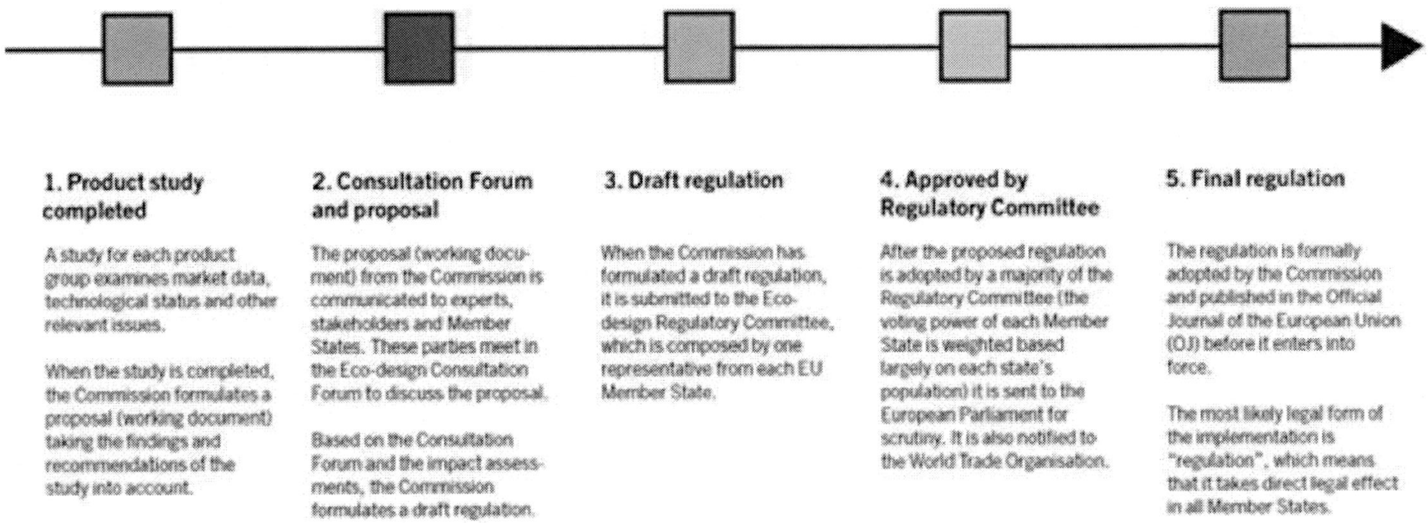

Source: ECEEE, 2011. http://www.eceee.org/Ecodesign/products.

Figure 11. Ecodesign Regulatory Process.

The final Ecodesign implementing measure must meet the following criteria:

- No significant negative impact on a product's functionality
- Health, safety and the environment must not be adversely affected
- No significant negative impact on consumers
- No significant negative impact on industry's competitiveness
- Does not impose proprietary technology on manufacturers
- No excessive administrative burden on manufacturers

3.4. Japan

Unlike the U.S., Australia and the EU, Japan's Top Runner standards differ from the typical minimum energy performance standards in that they are based on a maximum standard value system in which future targets are set using a base value equal to the most energy-efficient product on the market at the time of the value-setting process with an additional factor to take into consideration future technology improvement potential. Specifically, the Top Runner standard values are set as the highest efficiency current existing in the market plus consideration of the potential technological improvements for efficiency between the time of the value-setting and the target year. Once the target year is reached and the standard is in effect, manufacturers are considered in compliance if the average efficiency of all products sold, rather than the efficiency of every product sold, can meet the target standard. Thus, the target maximum standard value system provides more incentives for manufacturers to sell high efficiency product models in order to raise the average sales-weighted efficiency.

The Top Runner target standards are set by the Energy Efficiency Standards subcommittee under the METI Advisory Committee of Natural Resources and Energy. The standard setting process begins with an evaluation by the Energy Efficiency and Conservation Division of the Agency for Natural Resources to determine if a product is suitable for Top Runner program. To qualify, a product must have high sales volume in Japan, consume considerable amounts of energy while in use and requires significant effort to improve efficiency. At the same time, published Japanese guidelines exclude specialty equipment, equipment without established technical measurement and evaluation methods and equipment with extremely low market penetration from being considered for the Top Runner program. Products that meet the qualifying criteria are submitted in a proposal to the Energy Efficiency Standards subcommittee to be added to the Top Runner program. If the proposal is accepted, an Evaluation Standard Subcommittee is established for each product to evaluate specific standard values and factors. Working groups may also be formed under the Evaluation Standard Subcommittee to conduct studies about test procedures and measurement methods if none exist. The working group must also follow basic guidelines in setting the product scope and categorization. More specifically, basic indices of factors closely related to energy efficiency and factors representing consumer needs guide category development. Additive functions to product models should be disregarded in setting categories unless the additive function makes it very likely that all products cannot meet the standard value and will be forced out of the market. To incentivize the production of high efficiency models, efficient models with advanced technology and thus higher prices should included in the same category as other product models. Specialty goods, however, are excluded from this categorization but the technology behind their savings potential is considered when setting the target standard value.

Once the product scope and categories have been determined, the Subcommittee sets a proposed target value for each product category and a year for achieving the target value by working with industry to measure energy consumption for all products on the market and taking future technical progress into consideration. Depending on the product, the target standard value for each category can be set by a single numeric value or defined using a formula related to product attributes such as TV screen size. For home electric appliances and office equipment, the reduction of standby power consumption must be taken into account when setting the standard value. The target year, which differs by product, is set three to ten years ahead depending on the relationship between current efficiency levels, target value and outlook on technological progress.

After the target values and target year have been set, an interim report is made available for public comments before the draft standard is finalized and sent to the World Trade Organization to review for potential trade barrier concerns. The entire process for setting a new or revised Top Runner standard value generally ranges from 1 to 2.5 years, with the working group and evaluation studies lasting approximately 1 year and deliberations lasting half to 1 year.

After the standard is finalized, a commitment period begins during which manufacturers can work on improving the efficiency of their product and strive to meet the target standard. Once the commitment period ends, the target values become legally obligatory minimum sales-weighed energy performance standards and subsequent reviews and revisions are conducted by the regulator.

3.5. Comparison

There are key similarities in the standard setting criteria of standards programs in the four selected countries and regions, despite differences in each program's underlying regulatory framework and history. For all four programs, new or revised standards can only be set if it achieves significant energy savings through measures that are technologically feasible and economically justified. First, although the specific threshold for defining significant energy savings potential may differ between countries, the savings potential criterion is typically met if a given product has high sales volume and/or high per unit energy consumption. The EU Ecodesign directive for energy using products has a broader scope and covers other factors like waste, air pollutants and water usage in that it requires significant environmental impact, including energy consumption. Second, technological feasibility in all four programs are based on technical analysis of best available or beyond technology options for setting MEPS levels in U.S., Australia and the EU and analysis of potential technological improvement trends in setting the Top Runner target value and year in Japan. Third, key criteria for setting standard levels include ensuring that there are no excessive costs associated with a new or revised standard and economic impact analysis, particularly on consumers and manufacturers impact. Interestingly, though, Australia's Ministerial Council on Energy agreed in 2006 for the first time to consider regulating products even under circumstances where cost is imposed upon the community provided that regulation will offset even more expensive mitigation actions in the future (E3, 2009). Moreover, there is a similar timeframe on the order of around two years for developing and setting a new or revised standard in U.S., EU and Japan.

Besides overarching similarities in key features of the four standard setting processes, specific details of steps in the process differs. One example is in the types of analyses mandated in the standard setting process, which may range from a list of specific analyses that must be conducted in the case of the U.S. or broader evaluations conducted by product specific subcommittees in the case of Japan. Another example is in the definition of product classes for setting standard levels, which differs significantly between U.S. and Japan.

While U.S. explicitly differentiates product classes by their specific utility functions to consumers, such as separate class for refrigerators with the added feature of ice dispensers, Japan for the most part ignores products' additive functions in setting standards. This important difference in setting product classification has important implications for the allowable energy consumption, as the U.S. practice of separating product classes by additive functions that do not influence energy consumption will allow certain product classes to consume more energy than others and thereby reduce those manufacturer's incentives to improve energy efficiency. Japan, in contrast, explicitly aims to motivate manufacturers to improve energy efficiency through its product classification scheme and sales-weighted approach for reaching Top Runner target standards.

A summary of the major elements of each country's MEPS and energy labeling programs is shown in the table below.

4. Analytical Methods and Tools

With accurate and representative data as key inputs, well-founded technical, economic, scenario and cross-cutting analysis and tools serve as the basis for establishing minimum efficiency standards and labeling thresholds. While all countries have some analytical basis for determining the regulated efficiency levels, the scope and depth of these analyses and tools vary.

4.1. United States

As seen in Figure 5, DOE conducts many analyses in order to evaluate if it is meeting all the legislative criteria for establishing a new or revised MEPS. These analyses and the criteria they evaluate are shown in Table 2.

Screening Analysis

One of the first initial analyses conducted by DOE under its standards rulemaking is the screening analysis, a market and technology assessment used to identify product design options or efficiency levels that will be evaluated in the rulemaking.

Table 1. Comparison of International MEPS and Labeling Programs

	US MEPS	US ENERGY STAR	Australia	EU	Japan
Initial Legal Framework	1975 Energy Policy and Conservation Act	N/A - voluntary program	Created as a result of two 1980s state initiatives on energy labels, consolidated under national program in 1992 and first National Administrative Guidelines issued in 2000.	Directive 96/57/EC (MEPS) and Directive 92/57/ECC (labeling). 2008 Ecodesign Framework Directive.	Japanese Law Concerning the Ration Use of Energy (the Energy Conservation Law) of 1998
Regulatory Agency	Department of Energy	Dept. of Energy; Environmental Protection Agency	State and Territory governments, coordinated under the National Appliance and Equipment Energy Efficiency Committee; renamed Equipment Energy Efficiency (E3) Program	Implementation responsibilities belong to member states.	Energy Efficiency Standards subcommittee under the Ministry of Economy, Trade and Industry Advisory Committee of Natural Resources and Energy.
Standard Setting Principle	Achieve maximum efficiency that is technologically feasible and economically justified with significant energy savings possible.	Target of top 20% of efficient products on market. Most Efficient designated products may target ~top 5% efficient models in market.	Take into consideration world's current best regulatory practice MEPS in standard development process	Takes into consideration the best available technology and expected trends for technological development.	Base value set equal to the most energy-efficient product on the market at the time of the value-setting process. Target value takes into consideration future technological progress (i.e., may exceed current best available technology)

	US MEPS	**US ENERGY STAR**	**Australia**	**EU**	**Japan**
Standard Setting Timeline	Yes - timetable for test procedure and standards development and revision rulemaking set by legislation. A standards rulemaking typically takes around 3 years and regular rulemaking reviews are mandated for every 6 years.	Revisions conducted as necessary with market changes. No specific timelines are given.	Calendar for introducing new and revised MEPS/labels published every 3 years and supplemented with 10 year strategic plans for some broad end-uses (industrial equipment, gas equipment, commercial refrigeration). Review process typically launched 3 to 5 years after implementation.	MEPS must be reviewed four years after effective date; Label thresholds must be reviewed three years after label directive. Ecodesign preparatory studies take around 2 years.	Target year requiring target value be met by all manufacturers are typically set 3 to 10 years out depending on outlook of technological progress. Process for setting new or revised standard takes 1 to 2.5 years.
National versus Local Regulatory Precedence	Federal regulation has precedence, but states can adopt standards for non-covered products and/or petition Secretary of Energy for exemption	N/A – voluntary program	Programs enacted through state or territory legislation. It is intended that new national legislation will come into force during 2012 to replace existing regional legislation.	Prohibits member states from unilaterally adopting standards on traded goods.	Top Runner Standards are applied nationally.

This analysis identifies technology options that are viable for mass production in three to five years by using manufacturer websites, literature review and discussions with independent technical experts. The screening analysis also helps determine a price-efficiency relationship, though price may encompass factors besides cost. The screening analysis also evaluates a set of criteria in determining feasible design options or efficiency levels, including: technological feasibility; practicality to manufacture, install and service; adverse impacts on product functions or availability; and adverse impacts on health or safety (DOE, 2006).

Table 2. U.S. Standard-setting Criteria and Associated DOE Analyses

EPCA Criteria	Analysis DOE Performs to Address Factor	Rulemaking Stage
1. Economic impact of standard on consumers and manufacturers	Life-Cycle Cost (LCC) and Payback Analysis (including Markups)	ANOPR
	LCC Subgroup Analyses	NOPR
	Manufacturer Impact Analysis	NOPR
2. Lifetime operating cost savings resulting from standard	Life-Cycle Cost and Payback Analysis (including Markups)	ANOPR
3. Total projected energy savings resulting from standard	National Impact Analysis (including Shipments)	ANOPR
4. Impact of standard on utility or performance of products	Screening Analysis	ANOPR
	Engineering Analysis	ANOPR
5. Impact of any lessening of competition likely to result from standard	Manufacturer Impact Analysis	NOPR
6. Need for national energy conservation	National Impact Analysis (including Shipments)	ANOPR
7. Other factors the Secretary considers relevant	Environmental Assessment	NOPR
	Utility Impact Analysis	NOPR
	Employment Impact Analysis	NOPR
	Regulatory Impact Analysis	NOPR

Source: U.S. DOE 2006.

Engineering Analysis

The technologically feasible design options or efficiency levels identified form the screening analysis are further evaluated in the engineering analysis typically conducted by Navigant Consulting. This analysis is used to determine a cost-efficiency relationship which help evaluate which design change or efficiency level could save energy and to what degree. The analysis often involve purchasing models of different efficiency levels and dismantling it to itemize parts and costs, along with developing a model that accounts for investment costs outside of the cost for parts and labor (Rosenquist, 2010). This analysis is conducted in parallel with cost data provided by manufacturers.

Energy and Water Use Analysis

For each design option or efficiency level, the operational energy and water use is estimated based on usage patterns from RECS.

Mark-ups for Product Price Determination

The mark-up analysis is used to determine the mark-up and sales tax associated with converting a consumer price back to an estimated manufacturer cost based on census data and profit data from publicly traded companies.

Life-cycle Cost (LCC) and Payback Analysis

Since the economic impact of a standard on consumers is a major factor in the MEPS formulation process, the LCC and Payback analysis evaluates the life-cycle economic impacts of potential standard levels on consumers or end-users. Using inputs from the mark-up analysis; technical product data such as equipment lifetime, energy consumption, installation, maintenance and repair costs and estimates of future product and energy prices, a LCC and payback model is developed to calculate savings in operational costs over the product's life-cycle relative to any price increase related to adoption of a standard. Sensitivity analyses of discount rates and future energy price forecasts are also included in the analysis.

Shipment Analysis

The shipment analysis collects industry data on current shares of shipments by efficiency to feed into the National Impacts Analysis. This analysis is done by estimating current shipments and using models to forecast future shipments, which is fed into the Energy Information Administration's National Energy Modeling System (NEMS) tool for the shipment analysis.

National Impacts Analysis

This analysis is used to evaluate the potential energy and economic impacts associated with each design option or efficiency level at the national level. A spreadsheet-based accounting model of stock turnover analysis and forecasting models of U.S. residential and commercial energy use serve as the main tools for calculating a net present value of total consumer LCC and national energy and water savings. Under this analysis, a rebound effect is considered for energy use, but not economic analysis and future primary energy savings are also discounted. Results from the national impacts analysis serve as inputs for employment and environmental assessments.

Manufacturer Impact Analysis

A qualitative analysis of identified proposed standards' impact on manufacturers is conducted in three phases. First, an industry profile is created to characterize the industry with preliminary interviews with manufacturers to identify areas of concern. Second, manufacturers are interviewed with questionnaires to formulate the Government Regulatory Impact Model (GRIM) that helps assess industry and subgroup cash flow impacts and industry net present values. Based on the interviews and GRIM model, the impacts on competition, manufacturing capacity, employment and regulatory burden can be assessed.

LCC Subgroup Analysis

This analysis evaluates whether the proposed standards' impacts on consumers vary by region, demographic groups, or income levels in order to ensure that the standard does not disproportionately affect a certain subgroup of consumers or end-users. This analysis is conducted using the Monte Carlo probabilistic approach using nationally representative samples for different variables (energy prices, income, household size) from the Residential Energy Consumption Survey (more details in McMahon and Liu, 2000).

Employment Impact Analysis

This analysis evaluates the net jobs created or eliminated nationally amongst manufacturers, related service industries, energy suppliers and the economy in general by the proposed standards. This analysis is conducted using a national, 187-sector economic input/output econometric model called ImSET 3.1.1 developed by the Pacific Northwest National Laboratory. This model provides estimates of the change in national output for each sector based on data collected on initial investments, energy savings and economic activity associated with spending the savings resulting from standards.

Utility Impact Analysis

This analysis considers the impacts of potential standards on national electricity and gas suppliers using estimates of reduced energy sales, peak load and deferred power plant construction due to proposed efficiency standards.

This analysis is conducted using a version of the EIA NEMS tool, with annual energy savings from the National Energy Savings analysis as the model input. Each proposed standard level is compared to the Annual Energy Outlook's Reference Case to evaluate the amount of energy saved and its impact on utilities.

Regulatory Impact Analysis

This analysis evaluates and compares the national impacts of non-regulatory alternatives compared with proposed mandatory MEPS standards. The NEMS tool is also used to evaluate and compare the impact of non-regulatory alternatives to proposed MEPS standards.

Environmental Assessment

This assessment is conducted to determine potential reductions in the emissions of carbon dioxide and air pollutants of sulfur dioxide and nitrogen oxides associated with energy savings from the proposed standard levels. This analysis uses the same inputs and modeling tool as the Utility Impact Analysis, but with carbon and NO_x emissions as the key outputs of the analysis.

4.2. Australia

The main analytical method in support of MEPS program in Australia is the regulatory impact analysis that must be conducted before a product can be included in standards and labeling programs. The purpose of the regulatory impact analysis is to identify and compare the cost and benefits of each regulatory approach where benefits outweigh the costs across industry, consumers and regulators. The underlying basis for the regulatory impact statement and subsequent regulatory proposal include economic, engineering and statistical approaches of analysis, with consumer and industry research to inform analysis of consumer, industry, national and market impacts. The regulatory impact statement must include analysis of regional impacts and impacts of stakeholders likely to be most adversely effected, as specified by the Office of the Best Practice Regulation. The processes and steps in developing the regulatory impact statement and proposal are shown in Figure 7.

4.3. European Union

Similar to Australia, the EU conducts Ecodesign Preparatory Studies that serve as the basis for setting Ecodesign Implementing Measures on a country by country basis. These preparatory studies and related assessments are conducted by external experts and the European Commission. The first stage in the Ecodesign Preparatory Study involves defining a product and scope of the implementing measure, based on the listing of priority products for regulation. Once a product has been defined, economic and market analysis are conducted using generic trade and product data from the EU PRODCOM database and specific sales and stock data generated by specialist marketing sources to ensure the product has sufficient sales volume to meet the first criterion for Ecodesign regulation. Consumer behavior and local infrastructure are then evaluated in order to understand how the product is used and its end of life options.

In the following stages of the preparatory study, life-cycle analysis (LCA) is used to determine if the improvement potential is significant enough to warrant Ecodesign regulation using a base case assessment of existing products and an alternative case assessment of best available technology. The LCA follows a hybrid economic input-output and process-based LCA methodology based on the Energy-using Product EcoReport spreadsheet tool and data from the SimaPro database, which provides unit indicators in terms of fourteen environmental indicators such as energy, water, waste, global warming potential, and acidification per unit of material or process. A technical analysis of best available technology using the EcoReport tool helps identify potential design options for improvements and a ranking of the options is developed according to Least Life Cycle Cost (LLCC), which meets the third criterion of the Ecodesign Directive. The LLCC is calculated using a real discount rate based on European Central Bank data and a realistic product lifetime.

Finally, scenario, policy, impact and sensitivity analysis are conducted to evaluate the impact of different policy options and uncertainties surrounding the projected impacts. Scenario analysis from 1990 to 2020 is conducted to assess policy options other than Ecodirective implementing measures such as setting the best available technology as a promotional target, the LLCC option as the MEPS level, legislative or voluntary agreements and labeling. Each policy evaluation also includes impact analysis using consumer cost-benefit analysis, manufacturer impact analysis and assessments of impacts on competition, small firms, legal aid, sustainable development, carbon assessment, other environmental factors, health, race equality, gender equality, human rights and rural development (Defra, 2010b). Lastly, sensitivity analysis of all relevant factors including energy and resource prices, raw material and production costs and discount rates are included in the Ecodesign Preparatory Study.

4.4. Japan

There is no common analytical approach to setting the Top Runner standards and it appears that the methodology may vary by product as some products have a single numeric value for the target standard while others rely on a relational formula. Technical analysis is also part of the target setting process as previous trends in efficiency improvements and technological outlook are both considered when setting the target fiscal year and standard value for each product. A common tool used for market analysis and enforcement in Japan across Top Runner product classes is the product catalogues published biannually that reports all available models on the market and their energy performance.

4.5. Comparison

With the exception of Japan, the countries and region examined follow specific analytical methods and tools in the process for setting MEPS and Ecodesign implementing measures to ensure that regulatory criteria are met. For U.S., EU and Australia, consumer, manufacturer, national and regulatory impact analysis are all mandated in the standard setting process. In addition, the U.S. and EU also require the use of technical analysis to identify the best available technology options and life cycle cost analysis to evaluate the proposed regulatory level's impact on consumers. The EU stands out from the other three countries in that it adopts a life-cycle perspective of not only operational cost of the proposed Ecodesign implementing measure, but also of energy and environmental impact of the product.

Table 3. Analytical Tools and Methods used in International S&L Programs

	Analysis Overview	US	Australia	EU (Ecodesign)	Japan
Screening Analysis	screen potential product design options or efficiency levels	X	X	X	N/A
Engineering Analysis	evaluates and compare different design change or efficiency levels' effect on reducing energy use and cost	X	X	X	N/A
Energy and Water Use Analysis	estimate operational energy and water for efficiency level	X	X	X	N/A
Mark-up Analysis	Convert consumer price to estimated manufacturer cost	X	X		N/A
Life-cycle Cost and Payback Analysis	evaluates life-cycle economic impact of potential standard level on end-users	X	X	X	N/A
Market Analysis	evaluates efficiencies of current models in market		X	X	X
Shipment Analysis	current and forecast shipment analysis	X		X	X
National Impacts Analysis	evaluate potential energy and economic impact on national level	X	X	X	N/A
Manufacturer Impact Analysis	evaluates the impact on manufacturers' competitiveness, industry structure	X	X	X	N/A
Life-cycle Cost Subgroup Analysis	evaluate disparity of impacts on specific consumer groups	X	X	X	N/A
Employment Impact Analysis	evaluate net jobs created or eliminated	X	X	X	N/A
Utility Impact Analysis	evaluate impact on national electricity and gas suppliers	X	X		N/A
Regulatory Impact Analysis	evaluate and compare impacts of non-regulatory alternatives	X	X	X	N/A
Environmental Assessment	evaluate impact on CO_2, SO_2 and NO_x emissions	X	Only CO_2	X (lifecycle environmental impact)	N/A

5. DATA COLLECTION AND AVAILABILITY

Data collection and availability play important roles in the development and improvement of standards and labeling programs as well as program monitoring, verification and enforcement. Basic data on end-use usage patterns and energy consumption patterns can inform the standards development process by highlighting the major energy end-use consumers and their potential for energy savings. Similarly, sales data and data on the efficiency levels of products sold in the market can inform standards and labeling revision by illustrating the technical potential for efficiency improvement. Finally, a central and publicly accessible database on manufacturers' self-reported energy performance is an important monitoring, verification and enforcement tool for regulators and other stakeholders such as consumers, advocacy groups and other manufacturers.

5.1. United States

One of the major energy end-use and direct energy consumption data resources for informing standard and label development in the U.S. is the Residential Energy Consumption Survey (RECS) conducted by DOE's Energy Information Administration (EIA, 2011). This comprehensive household survey has been conducted every four years since 1978 and is publicly available, with the thirteenth RECS conducted most recently in 2009. RECS provide information on the physical characteristics of housing units, appliances usage, demographic and household characteristics, types of fuels used, and energy consumption and expenditure data for major fuels by collecting data from 4000 households that are statistically selected to represent all U.S. households. The data is collected through 45 minute in-person interviews with household occupants, questionnaires or phone interviews with rental agents and energy suppliers. The data is aggregated to represent four Census regions, nine census division and the four most populous states in the U.S.

In addition, data on market sales trends and efficiency levels are made available through the reporting requirements for ENERGY STAR retail partners. Specifically, retailers of ENERGY STAR-labeled clothes washers, dishwashers, air conditioners, and refrigerators are required to provide quarterly sales data to DOE on the types of product sold, total units in inventory, number of qualified units sold and model number, the store location and dates of sales. This data may also be made publicly available with the partner's identity masked. In addition, sales data and market trends are also collected by trade associations like Association of Home Appliance Manufacturers (AHAM) from its association members.

5.2. Australia

In Australia, key data inputs include national energy data collected by the Australia Bureau of Statistics as well as data from residential surveys and metering data (Ellis 2012). In 2009 to 2010, for instance, the Australia Bureau of Statistics conducted a Household Energy Expenditure Survey. In addition, local states and territories can also require the reporting of sales data specifically within its jurisdiction.. For consultation studies that evaluate whether products should be added to the MEPS and labeling programs, data is primarily purchased from market monitoring companies and covers products sold through retail channels and covered by import data. It does not cover industrial equipment that may be supplied directly from manufacturers or importers to users. There is, however, a national database of all registered products and their manufacturer-reported energy performance that is accessible to the public. The database with all registered models and their performance information can be combined with sales and import data serves as an important source for market trends such as sales-weighted average performance by product type.

5.3. European Union

Market surveillance is conducted by member states and there is no systematic collection of sales or energy consumption data on the regional level. The availability and ease of data collection will vary by country.

5.4. Japan

In Japan, sales market data is collected through cooperation with industry. Manufacturers' self-reported compliance data questionnaires are only intended for internal use and not shared with the public. In addition, manufacturers of products covered by the "Uniform Energy-Saving Label" have to register their product data with Japan's Energy Conservation Center in order to provide retailers with information to distribute the display labels.

5.5. Comparison

Of the four selected countries and region, the U.S. has the most robust data collection and availability for informing the development and revisions of MEPS and ENERGY STAR requirements with the regularly conducted RECS. RECS as a data resource is also important in that it is publicly available and can inform concerned stakeholders such as efficiency advocates or consumer groups as to which products may need to be added to the MEPS or ENERGY STAR program based on end-use usage patterns. In terms of sales data and efficiency trends of products on the market, all four regions take advantage of similar resources such as regional (e.g., EU member states or local Australian jurisdictions) or programmatic (e.g., ENERGY STAR, Top Runner, and Australia product registration) reporting requirements and purchased data from consulting companies and trade associations. U.S. and Australian sales and market data can be disclosed to the public, thereby making it possible for third-parties to cross-check and raise concerns about self-reported energy performance results. In contrast, only regulators can question compliance results in Japan because the Top Runner manufacturers' compliance data questionnaires are not disclosed to the public.

6. Stakeholder Participation

6.1. United States

The revised U.S. standard setting process was formulated with stakeholder and public participation as a priority and includes public comment periods of 30 to 60 days in which manufacturers, retailers, consumers, efficiency and environmental advocates, states and utilities are invited to provide input to milestones in the standard setting process. In addition, public meetings are also held after the initial analysis for standards setting is completed. The public meeting seeks the participation of one representative from trade associations, four to eight manufacturer representatives including international manufacturers, efficiency advocates from state agencies, environmental groups, non-governmental organizations and utility groups. Input and feedback from the public meetings is reviewed by DOE and addressed in the Notice of Proposed Rulemaking.

6.2. Australia

Stakeholder participation is incorporated into the Australian standards setting process through three main tracks: direct participation by relevant regional government agencies in the process through membership on the E3 committee; public participation in stakeholder forums held periodically throughout the year on specific topics; and through consultation processes built-in to the standards-making processes. As members of the E3 committee, officials from government agencies such as the Commonwealth, state and territory government agencies such as the Queensland department of Environment and Resource management, Mines and Energy, Western Australia's Energy Safety department, South Australia's Department for Transport, Energy and Infrastructure, Australian Capital Territory planning and land authority and New Zealand Energy Efficiency and Conservation Authority play direct role in the management of Australia's MEPS and labeling programs. Stakeholder forums for specific products, on the other hand, seek to bridge the E3 committee with broader spectrum of stakeholders by publicizing E3 plans related to MEPS and labeling, informing stakeholders of program achievements and milestones and provide opportunities for stakeholder feedback and input. In 2011, for example, there were six forums and conferences focused on product-specific S&L activities as well as compliance issues. In addition, for every proposed regulation, E3 is required to institute at least one consultation processes around the regulatory impact statement, and Standards Australia also consults stakeholders in developing or changing test methods and performance standards.

6.3. European Union

Similar to Australia, the EU also has an official protocol for stakeholder participation in the form of a Consultation Forum and informal public participation through stakeholder meetings for public input. The EU stakeholder meetings follow preparatory studies completed for Ecodesign implementing measures and provide opportunities for public input in the Ecodesign implementation process. The Consultation Forum was officially established under Article 18 of Directive 2005/32/EC to assess the economic, environmental and social impact of proposed implementing measures and adoption after completion of the preparatory studies. The Consultation Forum enables formal institutional participation of experts in contributing to the definition and review of implementing measures and monitoring of the effectiveness of existing mechanisms. It has up to 60 members including one representative from each member state and acceding country to the EU and one representative from 30 qualified organizations that may include small and medium enterprises, craft industry, trade unions, traders, retailers, importers, and environmental protection and consumer advocacy groups. In addition, the Consultation Forum is also open to observers from candidate and European Economic Area countries. The forum plays a crucial role in the Ecodesign implementing measure setting process since it reports directly to the Regulatory Committee which makes the final decision on implementing measures.

6.4. Japan

Stakeholders are directly included in the standards development process in Japan as they serve on the various committees that deliberate the technical details of potential standards, drafts and chooses the final standard and authorize the results of standard evaluations. The Advisory Committee for Natural Resources and Energy in charge of establishing and submitting standards and authorizing results is composed of members from industry, academia and research institutions, local governments and consumer organizations. Similarly, the Energy Efficiency Standards Committee that establishes and oversees the subcommittees responsible for preparing technical details of the draft standards are also made up of representatives from industry, academia, trade unions, consumer groups and other related corporations. The evaluation standard committee is also made up of stakeholders from the abovementioned groups but has meetings that are partially closed to the public in order to protect manufacturers' proprietary information. However, an interim report issued by the subcommittee is open for public comments.

6.5. Comparison

In all countries, stakeholders invited to participate in the different stages of standards development include industry/manufacturers, academic experts and consultants, trade associations, environmental and consumer advocates, and various levels of government officials. From the experiences of the four selected countries, the two key forms of stakeholder involvement and public participation are formal membership in committees and forums that inform the standard setting and regulatory decision-making processes and participation in informal stakeholder meetings or comment periods. All four countries are required to offer at least one open comment period for stakeholder input to the formulation of standards, with Japan, EU and Australia offering comment periods after an initial proposal or preparatory study for a standard is released.

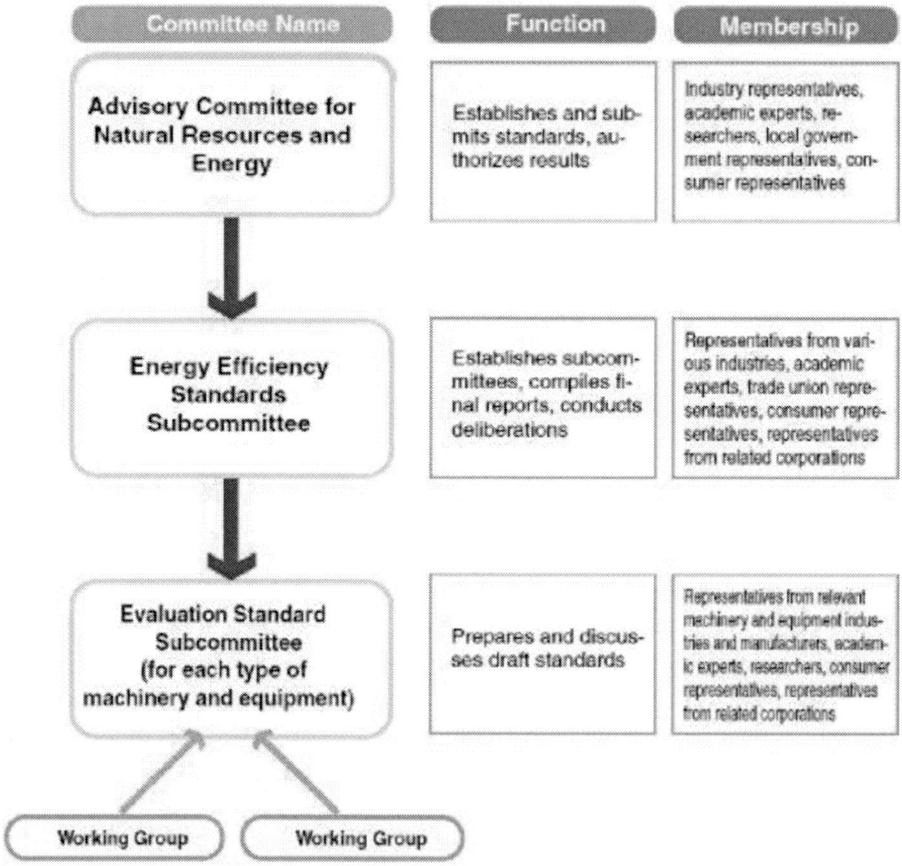

Source: Ministry of Economy, Trade and Industry (METI), 2010, "Top Runner Program: Developing the World's best Energy-Efficient Appliances (Revised edition March 2010)." Available at: http://www.enecho.meti.go.jp/policy/saveenergy/toprunner2010.03en.pdf.

Figure 12. Japanese Top Runner Standards Committee Structure and Stakeholder Involvement.

Table 4. Comparison of Stakeholder Participation in International S&L Programs

	Public Participation Opportunities	**Government /Regulators**	**Manufacturers /Industry**	**Consumer Advocacy Groups**	**Environmental Protection Groups**	**Power Utility Groups**	**Other Researchers (Academia, Consultants)**
US	Public meetings after initial analysis open to all; public comment periods	X	X	X	X	X	X
Australia	Product-specific stakeholder forums open to representatives from different groups	X	X	X	X	X	X
EU	Consultation Forum following Preparatory Studies open to representatives	X	X	X	X	X	X
Japan	Experts serve on Advisory Committee responsible for setting standard; Meetings partially closed to public	X	X	X		X	X

The U.S. stands out by requiring open comment periods during various stages of the standard setting process, including before any analysis is done during the product selection stage, and Australia also requires consultations with stakeholders during the standard-setting process. Of these four countries, the U.S. is the only country that does not grant formal membership to stakeholders but does require a lengthy comment period for all stakeholders to provide input throughout the standard setting process.

The EU, Japan and Standards Australia committees all ensure that key stakeholders are guaranteed a voice in the standard development and revision process with membership in regulatory committees responsible for setting and implementing the efficiency standards.

Table 4 illustrates the range of stakeholder participation in the standard setting and revision process of different countries.

7. Program Enforcement

7.1. United States

Enforcement of the U.S. MEPS program consists of two key components: certification testing and enforcement testing. Certification testing is a one-time process that occurs before a product can be sold in which the product manufacturer must certify via compliance statement and certification report that each basic model meets the MEPS requirement for that product class. The certification requirement applies to basic model, which is intended to streamline the certification process by grouping product models with essentially identical energy or water consumption-related characteristics but may differ in other irrelevant characteristics such as model color. Certification testing to ensure MEPS compliance may be conducted in-house or through an independent testing facility, except lighting and motors which must be tested in accredited labs from the National Institute of Standards and Technology's National Voluntary Laboratory Accreditation Program. The certified results report must include the following information: product type, class, manufacturer's name, model number and additional information specific to given product. The certified report is submitted to DOE via certified mail or email and annually to FTC for products covered by the EnergyGuide label. The civil penalty for knowingly violating MEPS standard is up to $110 for each violated product sold each day of non-compliance.

The second component of enforcement is verifying product compliance after certification through enforcement testing. Under the existing regulatory framework, the Secretary of Energy may request the initiation of enforcement testing only after receiving information in writing that a particular model by a particular manufacturer does not meet MEPS requirements. In the past, self policing amongst manufacturers and consumer rights advocacy groups were the main drivers behind initiating DOE enforcement testing of selected products. Once DOE receives written proof of non-compliance against a manufacturer, that manufacturer must supply a reasonable number (usually between 4 and 20) of sample units chosen at random by a DOE inspector for testing. Once selected, the sample units are boxed and must be shipped within five working days to a DOE designated lab for testing. If the units fail to meet the testing requirements, the manufacturer can request additional testing at their own cost by shipping more sample units and must cease distribution of that model in the meanwhile. If the model is still found to be out of compliance, the manufacturer must immediately cease all distribution and notify in writing all persons to whom it has sold the unit since the last date of compliance.

In September of 2010, DOE issued a notice of proposed rulemaking regarding changes to certification, compliance and enforcement for consumer products and commercial and industrial equipments.[2] The key change includes changing the one-time certification requirement to annual reporting requirement of test results for all models a manufacturer has in distribution for a given year, set to the FTC schedule for annual reporting requirements. Besides the reporting frequency requirement, the proposed rule would also expand coverage of information reported to include: manufacturer name, brand name, basic model number and individual model numbers, sample size, total number of tests performed and importer number from US Customs where applicable. The certification report will only be accepted via electronic submissions through the online Certification Compliance Management System and non-proprietary information in the certification report will be considered public and subject to disclosure. The proposed rule also establishes a record retention requirement in which manufacturers are required to keep certification test data and reports for all models being sold and for two years after sales have been discontinued. Finally, the proposed rulemaking seeks to develop a standardized process for seeking injunctive relief, civil penalties and other remedies against out of compliance manufacturers. In March 2011, the Final Rule was published which gave DOE the authority to proactively initiate enforcement testing without a written complaint, to test products from retail, distribution or manufacturer sources and provides an alternative approach to enforcement testing for manufacturers of low volume, customizable products.[3]

Besides proposed changes to MEPS enforcement, the US has also launched a verification testing pilot program in response to recent publications and a Government Accountability Office report raising concerns about the energy performance of certain ENERGY STAR-labeled products. This testing pilot program is also intended to support the State Energy Efficiency Appliance Rebate Program for ENERGY STAR-labeled products including refrigerators, freezers, dishwashers, clothes washers, gas tankless water heaters, gas storage water heaters and room air conditioners. Under this pilot program, 20% of basic models are randomly selected by sample units purchased from retailers and then tested at independent third-party laboratories following a standardized process. In Stage I testing, one sample unit is tested to verify performance within 5% of ENERGY STAR specifications. If the sample unit fails Stage I testing requirements, manufacturers are notified and given 10 days to request Stage II testing. Under Stage II testing, four to eight sample units will be tested to verify compliance with ENERGY STAR requirements. If a product is still found to be noncompliant after Stage II testing, the manufacturer is notified and given 20 days to respond with conclusive manufacturing or design evidence or quality assurance information on why ENERGY STAR requirements were not met. If the manufacturer does not respond or is not able to provide this information, DOE will refer the matter to the U.S. EPA for enforcement and notifies the states and public (including utilities' regional program sponsors, retailers and other stakeholders) of models being disqualified. As of August 2010, 41% of Stage I testing have been completed and results have indicated that of the 110 units tested thus far, 19 did not meet Stage I testing requirements. The distribution of Stage I testing results by product type is shown in Figure 13 below and the relatively high 17% non-compliance rate reflects the that enforcement and compliance testing has only recently received serious attention.

Lastly, trade associations such as AHAM and the Air Conditioning, Heating and Refrigeration Institute (AHRI) also administer voluntary certification programs that make important contributions to check-testing and certification.

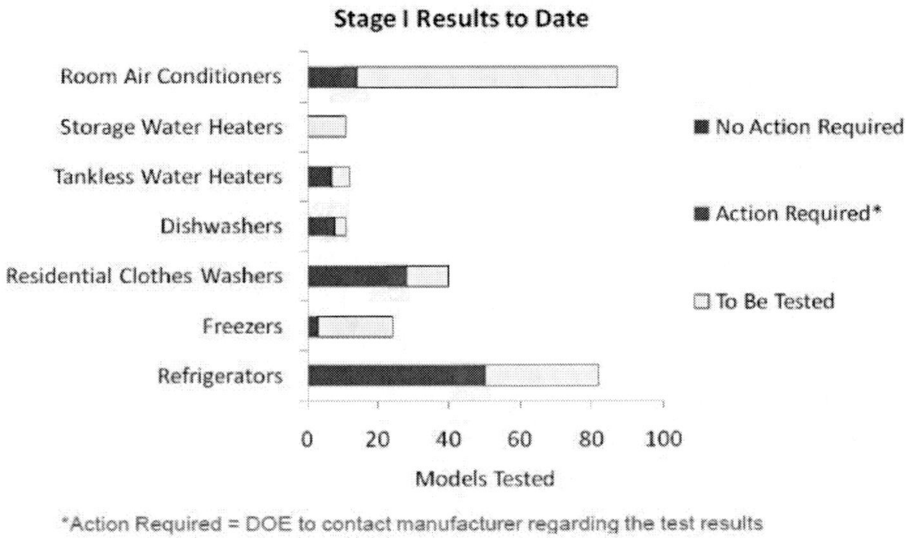

Source: taken from Karney, R., 2010, "Pilot Verification program for Selected ENERGY STAR Products." Department of Energy EERE Webinar on August 24, 2010. Available at: http://apps1.eere.energy.gov/buildings/publications/pdfs /corporate/ns/webinarENERGYSTAR testing 20100824.pdf.

Figure 13. Interim Results of U.S. ENERGY STAR Pilot Verification Testing Program.

7.2. Australia

The basis for enforcing Australia's MEPS and energy labeling requirements lies in the product registration programs administered in four states: New South Wales, Victoria, Queensland and South Australia. There are minor variations in registration fees, enforcement methods and penalties among the four registration programs, but a product registered in one state is recognized in all states and territories. To register its product model, a manufacturer submits an Application for Registration with information on the model description, supplier contact, claimed energy performance and a copy of the test results. The test data submitted with the product registration may be from non-accredited laboratories and may for some product categories be based on simulations or alternatives if testing of a physical sample unit is not feasible. However, whatever evidence is supplied, the supplier must attest that their product meets MEPS and/or the claimed performance for labeling. Data from the registration applications, with the exception of proprietary data, are placed in a user-searchable public register and updated daily. The registration database is intended for consumers to use and serves as a monitoring tool and compliance filter. Following a 2005 revision, product registrations are now active for four to five years, depending on the initial date of registration since registrations automatically expire on March 31st after three years of automatic renewal.

Australia has had a check-testing program since 1991 and reached milestone of 1000 check-tests completed in 2010. Check-testing is part of the National Greenhouse Strategy and about 25% of the E3 Committee budget is allocated for check-testing and round-robin testing with total testing costs of $500,000 in 2007-08 and $1.5 million in 2009-10. For product selection, the Australian check-testing program is risk-based and uses specific product and model selection criteria rather than randomly selected sample units. These criteria consist of the exclusion of products that were recently tested without any problems, selection that favors testing of newer models and brands, models with high volume of sales or higher self-claimed energy efficiency, models from suppliers with non-compliance record and if substantiated complaints were received from third parties such as other manufacturers, consumers or consume groups and other regulators.

Australia's check-testing program also consists of two stages of testing. In Stage 1 testing, a full or partial test is carried out following the given Australian Standard for 1 sample of the independently purchased unit by a laboratory accredited by Australia's National Association of Testing Authorities. Stage 1 testing costs are bore by the regulatory agency and E3. If a product fails Stage 1 testing, then a notice is sent to the registration holder (i.e., manufacturer or importer) informing them of intent to cancel the product registration. The registration holder then has a minimum of 15 days to make written requests to contest the registration cancellation by providing details and timetable for Stage 2 check-testing. For the product registration to remain active, the holder must be able to provide satisfactory Stage 2 test reports for multiple units randomly selected by regulatory representatives. In terms of sample unit selection, Stage 1 test sample units are purchased anonymously from retailer or wholesale supplier, although manufacturers may be offered the opportunity to inspect selected units at the test laboratory at the discretion of the program administrator. For Stage 2 testing, which is paid for by the registration holder, the manufacturer is asked to supply sample units that are randomly selected by regulators.

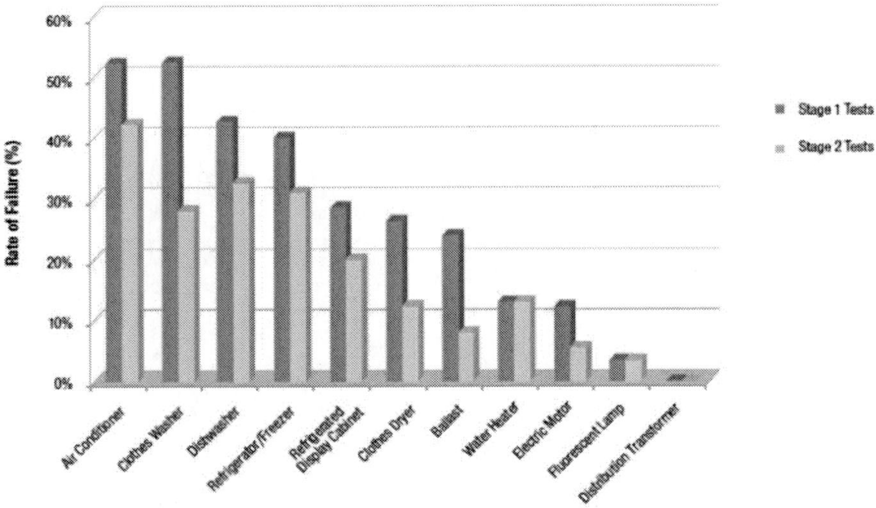

Source: Mark Ellis & Associates, 2011.

Figure 14. Australian Compliance Test Results for 1000 Check-tests by Stage of Testing, 1991-2010.

The Australian check-testing program may result in two types of non-compliance: energy noncompliance in which the tested product fails to comply with MEPS requirements and labeling noncompliance in which testing shows the product differs from self-claimed information, or if the product has a missing or non-conforming energy label. If a product is found to be in non-compliance, the local regulatory agency will cancel its registration and immediately notify agencies in other jurisdictions. cancelled product cannot be sold until alterations have been made to meet MEPS or labeling requirements and the product is re-registered. The results of the check-testing are all published, with those found to be selling non-compliant products publicly named to raise the perception that noncompliance is likely to be detected and action taken. In addition, through cooperation with the Competition and Consumer Commission, enforcement can also take the form of sanctions and fines imposed on companies for misleading and deceptive conduct for selling wrong-labeled or noncompliance equipment (Wiel and McMahon, 2005). This was exemplified in the case of LG, which was required to compensate eligible consumers with rebates ranging from $71 to $436 per unit for the additional electricity consumed by mislabeled air conditioners in 2006 (ACCC, 2006). National MEPS label compliance check-testing results are presented in Figure

14. Australian Compliance Test Results for 1000 Check-tests by Stage of Testing, 1991-2010.

Labeling display compliance is regularly checked by product type through surveys in retail and other outlets, including the internet.

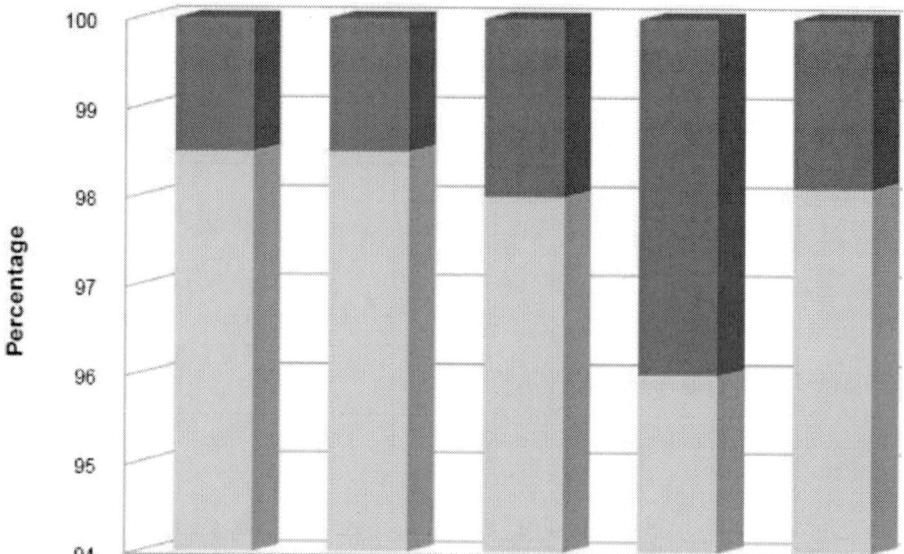

	Refrigerator/ Freezer	Clothes Washer	Clothes Dryer	Dishwasher	Total
■ Unlabelled	1.5%	1.5%	2.0%	4.0%	1.9%
Labelled	98.5%	98.5%	98.0%	96.0%	98.1%

Source: Australian Refrigeration Council Ltd., 2009.

Figure 15. Label Results of Australian Energy Label Compliance Surveys, 1998-2005.

Despite improvements in its check-testing programs as revealed by testing results in Figure 14. Australian Compliance Test Results for 1000 Check-tests by Stage of Testing, 1991-2010 and the Australian check-testing program still faces some challenges that are beginning to be addressed by program changes. Current regulatory language states that non-compliant products cannot be sold or supplied to others but can be used, thereby leading some companies to import non-compliant products of high values like transformers for self-use. Moreover, enforcement often varies by jurisdiction and is dependent on local resource availability and government prioritization.

7.3. European Union

Although the EU does not directly undertake compliance verification activities on a regional level, the MEPS and Energy Labeling Directives do outline test report documenting and recordkeeping requirements. For MEPS, Directive 96/57/EC specifies that the manufacturer's product test report must include the manufacturer's name and address, description of the model for identification, information on main design features or items that affect electricity use, operating instructions, and the results of test and details on conformity to MEPS requirements. The manufacturer or authorized representative within the European Community must keep test reports for at least three years from the date on which the last appliance was manufactured and the records are subjected to inspections by national authorities. Similarly, Directive 92/57/EC also requires that manufacturers have labeling documentation which includes a general description of the product, results of design calculations for determining the energy consumption level and test reports where available. The label documentation must also be available for inspection for up to five years after the last unit was manufactured.

Since the Ecodesign framework and implementing measures are still being developed on a product by product basis, there are no overarching compliance documentation or verification requirements. Rather, Directive 2005/32/EC states that it is the member states' responsibility to determine penalties for noncompliance and delegate market surveillance and enforcement responsibilities to the appropriate national authorities. In addition, the Directive states that national surveillance authorities should exchange information and "make the utmost use of electronic means of communication." If a product model is prohibited or withdrawn from the market due to non-compliance, the European Commission and other member states must be informed immediately.

In the example of the UK, the National Measurement Office (NMO) was recently appointed the Enforcement Authority for the Energy labeling and Energy Using Products Framework Directives and survey and compliance testing have been conducted under Defra's supervision since 2004 (Defra, 2008). These periodic testing initiatives are conducted for priority product groups, taking into consideration the level of policy support or participation in existing schemes. The selected test models will be based on market analysis, with aims to cover most manufacturers, new brands or a particular market sector. The sample units will be purchased anonymously from retailers, tested, and initial test results shared only with manufacturers. The manufacturers will then be given time of around four weeks to respond to the test results and explain any discrepancies. If a tested unit is not in compliance, then the manufacturer will be asked to repeat testing at an accredited testing laboratory for three additional samples for inclusion in the report. After meetings with suppliers and verification of the validity of the test results, a summary report will be published and publicized with full detailed test results by brand. Recent reviews of testing reports have found that manufacturer non-compliance rate for meeting the claimed energy level on the Energy Label is estimated to be around 10% to 15% with labeling display non-compliance rate of 20% for products without a correct label at the retail level (Defra, 2010d).

In addition, consultation proceedings are currently underway for evaluating potential penalty regimes and cost sharing arrangements. Since only criminal sanctions with a maximum fine of £5000 can be used to dissuade non-compliance, a 2010 draft amendment proposes introducing civil sanctions against noncompliance. In its consultation on introducing civil sanctions and cost sharing for implementing Ecodesign measures, the proposed UK amendment asserts that these civil sanctions should be "effective, proportionate and dissuasive" and take into account the extent of non-compliance in the European Community market (Defra, 2010c). The proposed sanction process begins with a Compliance Notice, which is issued to the manufacturer when the authority is satisfied beyond reasonable doubt that an offense against the Ecodesign implementing measures has been committed. After the notice of intent has been issued and the 28 days response period has passed, the NMO can enforce an unlimited variable monetary penalty. This final notice of intent to issue a sanction can be appealed if manufacturer can prove that the decision was based on error or wrong in law, if the penalty amount is unreasonable or other reasons. The enforcing authority may also issue a stop notice to prohibit the continuation of any activity that is deemed an offense, and can only be reversed if a completion certificate is issued to the manufacturer for taking corrective measures. If manufacturers fail to comply with the compliance notice, stop notice or other enforcement undertakings, an *additional* monetary penalty can be issued. Importers are also responsible for verifying compliance from imported models and for making technical compliance documentation available.

UK's implementation and compliance testing efforts are not representative of the EU and recent reviews of enforcement activities amongst the EU-15 member states have shown a range of enforcement efforts. In testing appliances for MEPS compliance, three out of nine original member states did not test appliances and only Denmark and the Netherlands performed many tests and reported the results centrally for enforcement action (Table 5). More recently, of all the EU member countries, only 17 countries have accredited test labs and of those, only seven countries have laboratories capable of conducting verification testing for more than one product. As a result, only 800 to 1400 product energy performance tests are conducted annually in the EU (Waide 2011). In some cases, retailers and consumer associations may also conduct their own third-party testing to verify the energy performance of products being sold. For non-compliance, penalties primarily consist of fines, which average €41,000 where reported but may be as low as €320, and negative publicity and publicized list of non-compliance producers (Waide 2011). In some cases, supply chain management may require third-party verification as a pre-requisite to stocking a particular product model.

Table 5. Range of Testing Activities in Selected EU Member States

Member State	Tests in 2005	Reported enforcement action
Finland	Few (< 10) appliances tested	No enforcement actions
Denmark**	Comprehensive tests (63 appliances and 10 A bulbs)	Reimbursement of costs of testing and handling
Germany	No testing*	No enforcement actions reported
United Kingdom	Few tests for enforcement*. Many tests by MTP and EST	No enforcement actions***
The Netherlands**	Comprehensive tests (100)	Compliance letters issued
France	No testing	Not reported
Austria**	No testing	No enforcement actions
Italy**	Few tests	Not reported
Greece	No testing	No enforcement actions
New Member States: - Czech Republic - Bulgaria - Poland - Lithuania - Romania	No testing	No enforcement actions

Table 4: Current status on test of appliances in interviewed EU Member States.

*Tests are carried out by local authorities with no central reporting

**Consumer organisations carry out additional tests, but not always according the EU test standards.

***Presently (2006), enforcement procedures are tested for six appliances.

Source: ANEC and UK Defra, 2007, "A review of the range of activity throughout Member States related to compliance with the EU Energy Label regulations in those countries." Report ANEC-R&T-2006-ENV-006. Available at: http://www.anec.org/attachments/ANEC-R&T-2006-ENV-008%20(final).pdf

Table 6. Range of Activities on Label Display Enforcement in Selected EU Member States

Member State	Shop inspections 2005	Reported enforcement action
Finland	250 shops inspected	Compliance letters issued
Denmark	100 shops inspected	Compliance letters issued Control visits carried out Police reports filed
Germany	Yes – not reported centrally*	Not reported centrally
United Kingdom	Yes – not reported centrally*	Not reported centrally
The Netherlands	700 shops inspected	Compliance letters issued
France	A number of shops inspected through a study**	No actions
Austria	Yes – not reported	Compliance letters issued
Italy	Yes – not reported**	Not reported
Greece	Yes – report to be presented	Not reported
New Member States: - Czech Republic - Bulgaria - Poland - Lithuania - Romania	Inspections initiated in Bulgaria and Romania	Not reported

Table 3: Current status on display of labels in interviewed EU Member States.
* Shop inspections are carried out by local authorities with no central reporting
** Additionally, 30 shops inspections have been carried out by the consumer organisations CLCV (France) and ADICONSUM (Italy) in a project covering France, Italy and Spain[9].

Source: ANEC and UK Defra, 2007.

Prior to the Ecodesign Directive, information sharing among energy authorities and related organizations were very low in the EU and even on a national level in some states where labeling implementation is decentralized to local authorities (Table 6). Currently, across the 30 member states of the European Economic Area, 80 full-time equivalent staff is estimated to work on Ecodesign and Energy Labeling compliance with a similar level of staff supporting store inspections of compliance with labeling directives (Waide 2011). In terms of financial resources, it is estimated that total expenditure on energy efficiency standards and labeling regulatory compliance is about €7 million per year across the entire EU region (Waide 2011).

7.4. Japan

Japan's enforcement of its Top Runner program differs from other countries since Top Runner is not a MEPS program, but rather based on a maximum standard value that can achieved on a sales-weighted basis.

Compliance and verification testing cannot be used to evaluate compliance with the Top Runner target standard since achievement of the target is measured by a sales-weighted average, not a per unit, efficiency of product models sold by a manufacturer. Instead, verification of Top Runner target standard achievement is completed using questionnaires distributed by the Agency for Natural Resources and Energy to all manufacturers after the target fiscal year has ended. These questionnaires collect information on the total number of units shipped and the energy efficiency of the units. Product catalogues with product information along with retail store surveys are periodically and continuously collected to confirm labeling display implementation and to validate the manufacturers' completed questionnaires.

In the event that a manufacturer is not able to meet the Top Runner target standard after the target year, there are several options for addressing non-compliance. METI can make recommendations to the manufacturer on improving their model's average energy efficiency. If these recommendations are not followed, Japan has traditionally relied on a "name and shame" approach in which manufacturers are pressured to comply after METI's recommendations and the name of the manufacturer are made public. In some cases, manufacturers may be ordered to adopt METI's recommendations and in the most extreme cases, a penalty of less than one million yen may be imposed for non-compliance.

There are, however, some caveats to the enforcement of the Top Runner program as a result of the composition of the Japanese manufacturing industry. Because the Japanese market is largely dominated by five to ten large manufacturers, the Top Runner program targets these large manufacturers and essentially exempts the much smaller manufacturers for competitive reasons. In particular, only manufacturers whose efficiency improvements will have substantial impact on energy consumption and whose organizational capacity is economically and financially stable will be subject to recommendations for improvements as part of the Top Runner program. Since enforcement of the Top Runner program for small manufacturers will have very small incremental energy impact on the overall efficiency levels, smaller firms are not subjected to strict enforcement and verification of their progress in achieving the Top Runner targets. In addition, if an entire category of products fails to meet the Top Runner targets, then an evaluation of why the target was not met, other companies' achievement records and other factors will be undertaken before compliance can be enforced.

7.5. Comparison

Although implementation mechanisms such as certification and manufacturer reporting requirements were included in the regulatory framework for all four countries, the extent and form of enforcement and compliance verification mechanisms differ significantly among the four regions examined (Table 7). All four regions have certification or registration requirements, with the U.S. and EU outlining specific reporting requirements in its standards and labeling regulations. Australia differs in that its product registration program is approved and managed by local jurisdictions, although the online registration system has increased the centralization of certification and registration in the absence of a national program. Japan does not have specific certification requirements, and requires only annual reports of sales and efficiency by model units from manufacturers.

Compared to the U.S. and Japan, the EU and Australia both have relatively established and extensive check-testing programs for enforcing compliance with both energy performance and labeling requirements, although specific enforcement efforts vary by jurisdiction (i.e., Australian states and territories and EU member states). Australia and the EU are also the only two regions that have a specific testing budget within its standards and labeling programs on the order of USD$0.5 to over USD$2 million USD. Both regions also have financial penalties for non-compliance, with the UK having criminal sanctions with maximum fines of up to £5000 and proposals for further civil sanctions and Australia's states and territories having financial penalties in addition to cancelling a product's registration.

The new U.S. pilot program initiated check-testing for selected ENERGY STAR-labeled products and follows the structure of the Australian program with its two stages of testing that allows manufacturers to seek additional testing if samples units fail the first stage of testing. However, the U.S. program differs in that it only tests for energy performance and does not include inspections on labeling compliance with the EnergyGuide or ENERGY STAR label. Japan does not have any formal check-testing or compliance testing programs and relies mostly on periodic reviews of product catalogues and retail store surveys for verification. Since neither the U.S. nor Japan has established testing programs, they rely heavily on informal enforcement mechanisms. Prior to the recent proposed revisions, the U.S. DOE conducted testing only if it received a written complaint against a manufacturer's product from a third party, such as competing manufacturers or consumer groups. The U.S. does have legal provisions for fines of up to

USD$110 per product per day of non-compliance. Similarly, the key approach to rectifying non-compliance in Japan is informally naming and shaming the manufacturer in public although fines of up to ¥1,000,000 are possible.

Table 7. Comparison of Major Elements of Standards and Labeling Enforcement Programs

	U.S.	Australia	EU/UK	Japan
Certification requirements	Previously one-time, now annual reporting requirements	Mandatory registration program for products	Documentation requirements for MEPS and Label	Self reported sales and efficiency in annual questionnaires
Check-testing	Pilot program for selected ENERGY STAR- labeled products started in 2010	Longest and most extensive check-testing program	Varies on a country- by-country basis	None; only inspections of product catalogues and retail store surveys
Sample Selection Method	Testing for reported non-compliance: DOE inspector selects from samples provided by manufacturer. ENERGY STAR pilot testing samples purchased from retailers.	Test sample models selected base on risk of failure and likely impact on program outcomes (e.g., newer, high volume or high claimed efficiency models). Stage 1 unit purchased anonymously from retailer or wholesale supplier. Stage 2 units randomly selected by regulators from samples provided by manufacturers.	Stage 1 units purchased anonymously from retailer. Stage 2 units provided by manufacturer.	None
Sample Size	1 unit for Stage 1, 4-8 units for Stage 2	1 unit for Stage 1, 2 or more units for Stage 2	1 unit for Stage 1, 3 units for Stage 2.	None
Testing Process	For both certification and ENERGY STAR check-testing, Stage 2 testing of products found to be noncompliant in Stage 1 can be requested by manufacturers	Two stages of testing, with Stage 2 requested by manufacturer.	Two stages of testing.	None

Table 7. (Continued)

	U.S.	Australia	EU/UK	Japan
Compliance verification	Energy testing only	Both MEPS and labeling compliance verified	MEPS and labeling compliance verified but varies by country	None
Fines/penalty for Non-compliance	USD$110/product/day	Product registration cancelled. On the spot fines and compensation for consumers have been negotiated in previous cases of non-compliance.	Criminal sanctions with maximum fine of £5000; proposal for civil sanctions	Up to ¥1,000,000
Testing budget	Not known	AUD $0.5-AUD $1.5 million dollars	Varies by country, UK has specific budget of around £0.6 to £1.9 million.	Not known
Informal enforcement mechanisms	Self-policing amongst manufacturers through complaints hotline	Public reporting of all compliance and enforcement activities and results, including the identification of suppliers of non-compliant products	None	Name and shame approach of publicizing non-compliance
Information sharing between agencies or jurisdictions	Very limited; DOE and EPA on ENERGY STAR	Between local jurisdictions and the Commonwealth	Ecodesign directive requires immediate information exchange b/w member states	None
Voluntary certification programs	AHAM; ASHRI certification programs open to all manufacturers	None	Eurovent; national promotional campaigns in UK and Denmark	None

Another important element of standards and labeling compliance enforcement is the extent to which there is information sharing between enforcement agencies or jurisdictions since appliances are often traded across state or regional borders. For decentralized programs such as in Australia and the EU, a product model that is found to be out of compliance in one state can easily be sold in another state if its poor performance is not detected, thereby undermining the effectiveness of regional enforcement. This is especially true in cases where some jurisdictions have more resources and capacity for enforcement than others. In recognizing this potential weakness, both Australia and the EU regulations call for immediate notification of other jurisdictions and agencies when a product is disqualified for noncompliance in one jurisdiction. This practice is less common in the U.S., where there is limited information sharing between DOE and EPA on ENERGY STAR compliance and a proposal for data sharing between the FTC and DOE on EnergyGuide products, and nonexistent in Japan. Lastly, U.S. and the EU have voluntary certification programs established by trade associations and public awareness campaigns that provide additional support to testing and enforcement.

8. Basis for Test Procedures

A consistent test standard for each product is an important foundation for MEPS and energy labels as it provides all manufacturers with a standard metric, a standard testing facility and standard test procedure and process for ensuring compliance with testing requirements (Wiel and McMahon, 2005). While countries may have different testing facility requirements such as national accreditation for different stages of the process (e.g., product certification versus enforcement testing), there are generally common characteristics in an ideal test procedure, including: repeatability (provides same result each time the same product is tested at the same lab), reproducibility (provides same results each time the same product is tested in different labs), accurate measure of energy consumption that reflects in-situ consumption, accurate measure of energy efficiency that reflects in-situ energy efficiency ranking and is not too expensive or time consuming to develop (Waide, 2010).

In many cases, countries already have established national industrial or product test standards that measures safety and performance and may include energy measurements. In other cases, countries partially or fully adopt and apply international standards developed by voluntary technical standardization organizations such as the International Standard Organization (ISO) and International Electrotechnical Commission (IEC). If national conditions are conducive to the adoption of international standards, then harmonizing national test procedures with international test standards provide advantages in facilitating cross-country trade, providing comparable benchmarks for quality, health & safety and the environment, and reducing the efforts needed for developing countries to develop a test procedure for new products (Waide, 2010). At present, however, the extent of international harmonization varies significantly between countries and products, with many countries having different energy test procedures and energy efficiency metrics.

8.1. United States

In the U.S., energy test procedures are developed by DOE for MEPS and by either DOE or EPA for ENERGY STAR specification requirements. Historically, U.S. test procedures for MEPS and ENERGY STAR specifications have been based on related national standards developed by the American National Standards Institute (ANSI) or standards developed by trade associations such as AHAM for home appliances, American Society of Heating, Refrigerating an Air Conditioning Engineers (ASHRAE) for air conditioning, and National Electrical Manufacturers Association (NEMA) for motors. For example, the Energy Policy Act of 2005 explicitly calls for the adoption of national and industrial standards for distribution transformers, commercial pre-rinse spray valves, and vending machines. Although some ANSI test procedures are derived from international test procedures, the majority of existing US test procedures is not harmonized with international procedures (Waide, 2010). More recently, there is beginning to be a shift towards standard harmonization in the U.S., as signaled by EISA 2007 mandating that test procedures for all residential product standards adopted after July 1, 2010 and other covered products be amended to include standby and off mode energy consumption, "taking into consideration the most current version of Standards 62301 and 62087 of the International Electrotechnical Commission"(EISA, 2007). In other words, the IEC standard for measuring standby and off mode energy consumption shall be adopted in the U.S. unless the current U.S. test procedure already covers these two modes or an integrated test procedure using the IEC standard is not technically feasible. Most recently in 2012, the final rule adopted for revising the refrigerator-freezers efficiency test procedure included new compartment temperatures and new methods for measuring compartment volumes in order to improve harmonization with existing international standards and test repeatability (US DOE, 2012).

8.2. Australia

Prior to the 1990s, test standards for MEPS and energy labeling programs in Australia were incorporated into State and Territory legislation and regulations. Recognizing the burden and cost of amending nonuniform regulations and technical details of the standards, a uniform approach was adopted by Standards Australia in which technical details of test procedures including test method, ambient conditions, performance measures (e.g., temperature operation test for refrigerators, minimum wash performance) and test materials were designated in Part 1 of Australian Standards and regulatory requirements for MEPS and labeling were designated in Part 2. Part 2 also include data on how to calculate energy label ratings, technical details on number of units to be tested, MEPS and labeling requirements and check testing procedures. Both parts of Australian Standards are drafted by relevant standards committee composed of representatives from State and Territory regulatory agencies, E3 Committee, industry, consumer and other relevant stakeholders and published by the national standardization body, Standards Australia. Part 2 of the standard must be unanimously approved by relevant State and Territory regulatory agencies before it can be published as Australian Standards.

In developing and revising test standards, Standards Australia follows four guiding principles, including that standards will benefit the Australian community, Australia will influence the development of and maximize use of relevant International Standards, standards development will be driven by the needs and relies on commitment of stakeholders and that Australian Standards will only be produced where appropriate (Standards Australia, 2010). As the second principle highlights, Australia emphasizes the need for international harmonization and asserts that Australian or Joint Australia/New Zealand Standards should adopt relevant IEC and ISO test standards where feasible. The adopted test methods must also be repeatable and reproducible, cater to every design on Australian and New Zealand markets and aim to be relevant to actual usage by consumers (NAEEEP, 2005). In fact, Australia requires that manufacturers state explicitly parameters such as capacity and settings used in performance tests so consumers can effectively choose an energy efficient product that meets their specific needs.

8.3. European Union

With the establishment of a single common European market that emphasizes the removal of trade barriers, EU policy has been to adopt and use international test procedures whenever they are available and sufficient for regulatory purposes. In fact, the EU has created EU-level shadow standard bodies to the ISO and IEC with the creation of European Committee for Standardization (CEN) and the European Committee for Electrotechnical Standardization (CENELEC), respectively. Both CEN and CENELEC frequently adopt ISO and IEC standards, but may modify them to reflect European conditions and in rare cases, adopt entirely different standards from another country or develop a dedicated European national standard. ISO and IEC may also sometimes adopt or adapt CEN or CENELEC standards for certain products (Waide, 2010). For instance, seven of the eight products subjected to the EU energy label are tested using European test procedures that are essentially equivalent to corresponding IEC or ISO test procedures while the remaining product does not have an applicable international test procedure (Waide, 2010). Similarly, seven of the eight products subject to MEPS under the Ecodesign directive uses internationally aligned energy test procedures.

8.4. Japan

One of the guiding principles for the Top Runner program is to harmonize with international test procedures and existing domestic test procedures when possible. This is reflected in Principle 10, which states:

> "Measurement methods should bear domestic and international harmonization in mind. If a standard has been already established, the measurement method should harmonize with the standard to the extent possible. Where no measurement method standard exists, it is appropriate to adopt specific, objective and quantitative measurement methods based on actual equipment usage (METI, 2010)."

This principle further calls for harmonization with existing voluntary or compulsory standards, including ISO/IEC standards and Japanese Industrial Standards. If no international or domestic standard exists, then Japan will adopt a national test procedure or consider using another country's test procedure. Although it is a guiding principle, harmonization is not a binding objective and Japan's existing test procedures reflect a mix of national and international test procedures. Nevertheless, even when differences exist, many aspects of national test procedures still correspond to elements in related international test procedures (Waide, 2010).

8.5. Comparison

While test procedures and methods found in the four selected countries and region vary in their degrees of international harmonization, a common guiding principle that has emerged over time is that test procedures should be harmonized to the extent possible while still reflecting national conditions and actual usage. This principle was only recently accepted in the U.S. with the 2007 shift towards harmonization of standby and off mode power consumption test methods with ISO/IEC test method. As a result, U.S. test procedures have historically been based on national industrial test methods and is the least harmonized of the four. This is likely because the U.S. standards program has historically been highly centralized on the federal level, and there was no need for standards harmonization to eliminate trade barriers between state jurisdictions (as in Australia) or countries (as in the EU). In contrast, the EU with its various member countries and Australia with its states and territories have emphasized the need for uniform test procedures and appears to have greater international harmonization. The countries also vary in the extent to which ideal elements of test procedures such as reproducibility and repeatability are integrated and emphasized in the basic procedures for developing and revising testing methods.

9. Program Resources

9.1. United States

Although the Energy Policy Act of 2005 allocated budget appropriations of up to USD $90 million per fiscal year through 2010, the actual program budget for appliance and commercial standards is smaller and split amongst different institutions. The appliance and commercial equipment standards group within the Building Technologies Program of DOE's Office Energy Efficiency & Renewable Energy (EERE) has an annual budget of USD $16 to USD $20 million. In terms of personnel, the EERE appliance program currently has 8 full-time staff members for developing and revising test procedures, including 1 staff member dedicated to certification and enforcement, and 8 staff members on standards analysis (EERE, 2009). In addition, Lawrence Berkeley National Laboratory (LBNL)'s Energy Efficiency Standards group and Navigant Consulting, two of DOE's primary contractors for technical assistance in the standard setting process, had funding of USD $4.3 million over a period of three years. Within the LBNL group, three to four full time staff members are assigned to work on a given product during the rulemaking process (Rosenquist, 2010).

Besides DOE's certification and enforcement program (discussed further in Section A.5.1), another important resource is the voluntary certification programs managed by two of the largest appliance manufacturer trade associations. Both AHRI and AHAM have voluntary certification programs that are open to its association members as well as non-members. Together, these two programs cover heating, ventilation, air conditioning, commercial and residential refrigeration, air conditioners, dehumidifiers and room air cleaners. For the AHRI program, manufacturers send in a test sample to an independent, third-party contracted testing laboratory to conduct and certify test results. For AHAM certification, manufacturers also need to undergo certification testing at an acceptable laboratory and the results are verified by AHAM at a third-party independent laboratory. Certified product models are listed in directories or databases managed by the trade associations and accessible to consumers. For the AHAM certification program, certified products may be randomly picked and check-tested by AHAM at an accredited laboratory.

9.2. Australia

Standards and labeling development and program management is funded by the E3 Program (formerly the NAEEC program), which was set up in 1992 and supported by program resources from each jurisdiction. Table 8 shows that the total estimated program budget in Australia has grown significantly in recent years, from AUD $5 million for 2006-7 to over AUD $10 million for 2010-11. The E3 receives 60% of its fund from the Commonwealth government, 20% from Australian states and territories (with individual shares proportional to population) and 20% from New Zealand. Funding from both the Commonwealth and States and Territories are from the National Framework for Energy Efficiency, this was set up by the Ministerial Council on Energy to advance energy efficient products in 2004, or in the form of staff and overhead. In 2008-2009, the NFEE provided AUD $2.4 million for the Australian standards and labeling programs. The largest source of support for the programs is actually from jurisdictional funding, which was on the order of nearly AUD $8 million in 2008-09. In particular, each jurisdiction must allocate staff and monetary resources to cover label registration, implementation and other program overhead costs (e.g., compliance testing and in-store compliance checks, technical support on product profiles and preparation of Regulatory Impact Statements). Although local jurisdictions receive some revenues from registration fees, which range from AUD $150 to AUD $285 by region, and sales of tested sample units to manufacturers for supplier test laboratory calibration, this makes up less than 5% of the total program budget.

Table 8. Estimated Australia national administrative resource commitment 2006/07 to 2010/11 MEPS and energy labeling

	$ Million				
	2006/07	2007/08	2008/09	2009/10	2010/11
Staff & overheads - Comm	2.2	3.2	6.8	4.8	6.5
NFEE Contributions – Comm	0.8	0.8	1.2	1.4	1.4
Total Comm	3.0	4.0	8.0	6.2	7.9
Staff & overheads - S&T (a)	1.1	1.1	1.1	1.1	1.1
NFEE Contributions - S&T	0.8	0.8	1.2	1.4	1.4
Total S&T	1.9	1.9	2.3	2.5	2.5
Registration Revenues (b)	0.45	0.45	0.45	0.63	0.63
Staff & overheads - Total	3.3	4.3	7.9	5.9	7.6
NFEE Contributions – Total (c)	1.6	1.6	2.4	2.8	2.8
Total Program	4.9	5.9	10.3	8.7	10.4
Net of registration revenues	4.5	5.5	9.9	8.1	9.8

Source: Personal communication from agencies. (a) Assumes S&T resource commitments were and will remain at 2008/09 level. (b) Allows for expected doubling of Vic revenues in 2009/10. (c) Excludes NZ contribution (20% of NFEE total)

Taken from: George Wilkenfeld and Associates and Marsden Jacob Associates, 2010, "Consultation Regulation Impact Statement: National Legislation for Appliance and Equipment Minimum Energy Performance Standards (MEPS) and Energy Labeling." Available at: http://www.energyrating.gov.au/library/pubs/201001-consultation-risnational-MEPS-labelling.pdf

In terms of personnel, the Commonwealth of Australia funds the majority of staff for the standards and labeling programs, with 28 full-time equivalent staff members while the jurisdictions fund a total of 9 full-time equivalent staff members (Table 9).

Table 9. Estimated Australia administrative resource commitment by jurisdiction 2008/09, MEPS and energy labeling

	$ '000											
	NSW	Vic	Qld	SA	WA	Tas	NT	ACT	State & Territ.	Comm	Total	Comm share
Staff and overheads	230	340	367	69	61	7	23	5	1101	6800	7901	86%
NFEE Contributions (a)									1200	1200	2400	50%
Total									2301	8000	10301	78%
Registration Revenue	191	180 (b)	62	13	0	0	0	0	446	0	446	0%
Share of registrations (c)	40%	33%	19%	2%								
Staff (FTE)	1.6	1.5	4.0	0.6	0.6	0.2	0.2	0.15	8.85	28	36.85	76%

Source: Personal communication from agencies. (a) Excludes NZ contribution (20% of NFEE total of $3M) (b) Projected to approximately double in 2009/10, with announced fee increase. (c) Oct 2008 to Oct 2009. 6% of registrations were in New Zealand

Taken from: Wilkenfeld and Jacob and Associates, 2010.

9.3. European Union

As the EU delegates all MEPS and labeling program responsibilities to its member states, program resource availability vary on a country-by-country basis. The UK, which has an extensively developed MEPS and labeling program, is used to illustrate, but not represent, the scale of national programs within the EU. The UK standards and labeling programs are included in its Market Transformation Programme, which had growing budgets of £1.65 million between 1998 and 2001 to annual budget of £3.3 million from 2006-07 (Chesshire, 2000 and Lloyd, 2007). This budget growth is in part due to new funding for the Market Transformation Program by new landfill tax receipts under the Business Resource Efficiency and Waste Programme. Most of the program budget is spent on direct expenses of staff salary and overhead expenses and the costs of lead and sub-contractors. The annual enforcement costs for the UK programs are estimated to range from £0.6 to £1.9 million.

In February 2010, the UK government conducted and drafted a cost-sharing impact assessment for compliance costs of its market surveillance framework for ensuring compliance with the Ecodesign implementing measures (Defra, 2010b). The proposal included two cost-sharing options, the first of which disregards compliance testing results with the government responsible for 25% of all costs and industry responsible for 75% of the costs if the result of the first of the four tests falls outside the tolerance zone. The second would require cost-sharing only for non-compliant products, as all tests are funded by the government but manufacturers must reimburse the government for testing costs if there is proof of non-compliance for one of its products. The impact assessment found that the second option is preferred with total annual testing costs assumed to reach £762,000, including annual costs of £162,000 borne by the industry under a scenario of 200 tests per year and industry paying for 27% of tests.

9.4. Japan

No specific information has been found on the specific financial or human resources for the Top Runner program. However, in 2002, Japan's total public budget for publicly financed energy efficiency measures in 2002 was ¥130 billion, or approximately €880 million assuming an exchange rate of €6.787 per ¥1000 (Nordqvist, 2006).

9.5. Comparison

Different levels of available information on budget availability and constraints as well as different scope and responsibilities make it difficult to directly compare national or regional standards and labeling programs, but some broad similarities and differences can be noted. Despite differing scopes, most of the programmatic budgets for the standards and labeling programs were in the range of USD$5 million to USD $20 million, assuming current currency exchange rates. Both Australia and the UK had budgets of around $10 million USD for their programs, with the UK budget including the Market Transformation Programme budget, annual enforcement costs and expected testing costs for the Ecodesign market surveillance framework. However, it must be noted that a different number of standards are developed and implemented in each country so the total programmatic budgets may not be directly comparable.

Australia is unique in that its programmatic budget is divided between the Commonwealth of Australia and its state and territories, with states and territories contributing as much as 25% of the programmatic budget. In contrast, the European Commission does not provide any direct financial support for the EU MEPS, labeling or Ecodesign policies and instead, places all implementation responsibilities on the member states. Although the UK program is well developed and funded, the lack of EU-level funding means that implementation and enforcement of standards and labeling will be highly dependent on the institutional support and financial capacity of individual member states (see Section A.5 for further details). The U.S. program is funded and implemented entirely by the federal government, with some states playing minor roles in administering complementary or supplementary programs such as promotional and customer awareness campaigns, ENERGY STAR program rebates and testing and compliance verification in the case of California (see CEC 2009 for more details on the California program).

Table 10. International S&L Program Resources

	Approximate Total Budget	**Funding Sources**	**Staff Resources**
US	~ USD $20 million/year	National Budget via Federal Government	~ 100 employees total (including contractors)
Australia	~ AUD $10 million/year	75% from Commonwealth government, 25% from states and territories	~ 40 full-time equivalent staff
EU (UK example)	~ £3.3 million/year	National government via Market Transformation Program; cost-sharing being considered for enforcement program	Unknown

Note: Approximate total budgets may not be directly comparable as the number of standards and labels being developed may differ between countries for a given year.

10. KEY FINDINGS AND CONCLUSION

The international review of standards and labeling programs in the U.S., Australia, EU and Japan has uncovered some overarching themes and highlighted several key factors to successful program elements. For example, standard-setting and programmatic implementation can benefit significantly from a legal framework that directly specifies a timeline or schedule for standard-setting and revision, product coverage and legal sanctions for non-compliance. For the specifics components of standard-setting, programs in the four countries revealed similarities in guiding principles that focus on achieving significant energy savings that are technically feasible and economically justified. In terms of analytical support for standard-setting, detailed survey data such as the U.S. Residential Energy Consumption Survey and rigorous analyses provide a strong foundation for setting a particular standard level. Similarly, the standard setting process can also be strengthened by involving different groups of stakeholders, while the particular form of participation may vary between countries. Sufficient program resources are critical to the effectiveness of standards and labeling programs and cost-sharing between national and local governments can be undertaken to ensure adequate resources and uniform implementation. Lastly, check-testing and punitive measures are important forms of enforcement while the cancellation of registration or product sales-based fines have also proven effective deterrents for non-compliance.

The international review of the four selected regions also illustrates the differing degree to which program development and implementation have been influenced by the level of government decentralization in a specific country. In addition, while no single country has best practices in all elements of standards and labeling development and implementation, national examples of best practices for individual elements do exist. For example, the U.S. has demonstrated rigorous analyses for standard-setting and robust data source with the RECS database while Japan's Top Runner standard-setting principle has been effective in motivating manufacturers to exceed targets ahead of time. In terms of standards implementation and enforcement, Australia has demonstrated success in enforcement with its long history of check-testing and enforcement initiatives while mandatory information-sharing between EU jurisdictions on compliance results is another important enforcement mechanism. As reflected by these examples, it is important to understand not only the drivers of different paths of standards and labeling development, but also the country-specific context for examples of best practices in understanding why certain S&L programs have been effective.

ACKNOWLEDGMENTS

The authors would like to thank Yukari Yamashita and her colleagues at the Institute of Energy Economics in Japan for their comments and review. The authors are also grateful to Mark Ellis for his invaluable comments and feedback on this report.

This work was supported by the China Sustainable Energy Program of the Energy Foundation and the Collaborative Labeling and Appliance Standards Program through the U.S. Department of Energy under Contract No. DE-AC02-05CH11231.

References

Air Conditioning, Heating and Refrigeration Institute (AHRI). 2010. "Certification." Available at: http://www.ahrinet.org/Content/AHRI CertificationPrograms732.aspx

ANEC (European Consumer Voice in Standardisation) and UK Department of Environment, Food and Rural Affairs. 2007. "A review of the range of activity throughout Member States related to compliance with the EU Energy Label regulations in those countries." Report ANEC-R&T-2006-ENV-006. Available at: http://www.anec.org/attachments/ANEC-R&T-2006-ENV-008%20(final).pdf

Association of Home Appliance Manufacturers. 2010. "AHAM Certification Program Information for Manufacturers." Available at: http://www.aham.org/industry/ht/d/sp/i/42721/pid/42721

Australian Competition and Consumer Commission (ACCC). 2006. "LG Compensates Consumers over Misleading Energy Ratings." Available at: http://www.accc.gov.au/content/index.phtml/itemId/763524/fromItemId /142

Australian Refrigeration Council Ltd., 2009. *Survey of Compliance with Energy Efficiency Labeling Laws: Whitegoods 2009*. http://www.energyrating.gov.au/wp-content/uploads/EnergyRatingDocuments /Library/Whitegoods/Whitegoods/200910-labellingcompliance-survey2.pdf

California Energy Commission (CEC). 2009. "2009 Appliance Efficiency Regulations." Sacramento, CA: Regulation CEC-400-2009-013. Available at: http://www.energy.ca.gov/2009publications/CEC-400-2009- 013/CEC-400-2009-013.PDF

Chesshire, J. H. 2000. "Independent Review of the Market Transformation Programme." Report prepared for the Department of Environment, Food and Rural Affairs. Available at: http://efficientproducts.defra.gov.uk/spm /files/download/byname/file/Chesshire Report.pdf

Commonwealth of Australia. 2008. "The National Greenhouse Gas Strategy." Available at: http://www.australianpolitics.com/foreign/environment/ ngs.pdf

Department for Environment, Food and Rural Affairs (Defra). 2008. "BNXS23: Procedure for Defra product monitoring checks, Version 3.1." Available at: http://efficient-products.defra.gov.uk/compliance

Defra. 2010a. "Consultation Annex 4: Draft Impact Assessment of the Proposed Penalty Regime for the Energy Using Products and Energy

Labeling Regulations." Available at: http://www.defra.gov.uk/corporate /consult/eup-labelling2010/20100323-consult-annex4.pdf

Defra. 2010b. "Consultation Annex 5: Draft Impact Assessment of Cost Sharing options available to the Market Surveillance and Authroity under the Energy Using Products and Energy Labeling Regulations."

Available at: http://www.defra.gov.uk/corporate/consult/eup-labelling2010 /20100323-consultannex5.pdf

Defra. 2010c. "Consultation Annex 2: Draft Ecodesign for Energy-Using Products (Amendment) (Civil Sanctions) Regulations 2010." Available at: http://www.defra.gov.uk/corporate/consult/euplabelling2010/20100323-consult-annex2.pdf

Defra. 2010d. "Defra MTP Compliance Strategy." Available at: http://efficientproducts.defra.gov.uk/compliance

Directive 92/75/EEC on the indication by labeling and standard product information of the consumption of energy and other resources by household appliances. Official Journal L 297, 13/10/1992. 22 September 1992.

Directive 96/57/EC on energy efficiency requirements for household electric refrigerators, freezers and combination thereof. Official Journal L 236, 18/09/1996. 3 September 1996.

Directive 205/32/EC establishing a framework for the setting of ecodesign requirements for energy-using products and amending Council Directive 92/42/EEC and Directives 96/57/EC and 2000/55/EC. Official Journal L 191, 22/07/2005. 6 July 2005.

Ellis, Mark. Personal Communication. 13 February 2012.

Equipment Energy Efficiency (E3), 2006, "The MEPS and Energy Labeling Process in Australia and New Zealand." Available at: www.energyrating.gov.au/pubs/meps-labelling-process-au-nz.pdf

E3. 2009. *Achievements 2007/08: Equipment Energy Efficiency Program Annual Report.* Barton: Commonwealth Copyright Administration.

E3. 2011. *Achievements 2009/10: Equipment Energy Efficiency Program Annual Report.* Report No: 2011/03. Barton: Commonwealth Copyright Administration.

Energy Information Administration, 2011, "Residential Energy Consumption Survey." Available at: http://www.eia.doe.gov/emeu/recs/

Energy Independence and Security Act (EISA) of 2007. Public Law 110-140. 110th Congress, 1st Session, 19 December 2007.

Energy Policy Act of 2005. Public Law 109-58. 109th Congress, 1st Session, 8 August 2005.

"Energy Conservation Program: Certification, Compliance, and Enforcement for Consumer Products and Commercial and Industrial Equipment." *Federal Register* 75 (Sept. 16, 2010).

"Establishment of the working plan for 2009-2011 under the Ecodesign Directive." Commission of the European Communities Communication COM 660. Brussels: 21 October 2008.

George Wilkenfeld and Associates and Marsden Jacob Associates. 2010. "Consultation Regulation Impact Statement: National Legislation for Appliance and Equipment Minimum Energy Performance Standards (MEPS) and Energy Labeling." Available at: www.energyrating. gov.au/.../details201001-risnational-MEPS-labelling. html

Government of Japan. 1998. "Energy Conservation Law of Japan (revised). Tokyo: Government of Japan.

Karney, R., 2010, "Pilot Verification program for Selected ENERGY STAR Products." Department of Energy EERE Webinar on August 24, 2010. Available at: http://apps1.eere.energy.gov/buildings/publications/pdfs /corporate/ns/webinar ENERGY STAR testing 20100824.pdf

Lloyd, R., 2007, "The Business Resource Efficiency and Waste (BREW) Programme." Department of Environment, Food and Rural Affairs presentation. Available at: www.crn.org.uk/system/files/content /documents/rebeccalloyd.pdf

Mark Ellis & Associates, 2011. *Performance Verified.* New South Wales: E3 Program Report. http://www.energyrating.gov.au/wp-content/uploads /Energy Rating Documents/Library/General/Compliance/201105-1000-checktests.pdf

McMahon, J. E. and X.M. Liu. 2000. "Variability of Consumer Impacts from Energy Efficiency Standards." Proceedings at the 2nd International Conference on Energy Efficiency in Household Appliances and Lighting, Naples, Italy, September 27 – 29, 2000. Available at: http://www.osti.gov/bridge/servlets/purl/779734-5qL5YU/native/779734.pdf

Ministry of Economy, Trade and Industry (METI). 2010. "Top Runner Program: Developing the World's best Energy-Efficient Appliances (Revised edition March 2010)." Available at: http://www.enecho.meti. go.jp/policy/saveenergy/toprunner2010.03en.pdf

National Appliance and Equipment Energy Efficiency Program (NAEEEP). 2005. "Administrative Guidelines for the Appliance and Equipment Energy Efficiency Program of Mandatory Labeling and Minimum Energy

Performance Standards Edition 5, June 2005." Available at: http://www.energyrating.gov.au/pubs/admin-guidelines.pdf

Rosenquist, Gregory. Personal Interview. 24 August 2010.

Standards Australia. 2010. "Australian Standards – Guiding Principles." Available at: http://www.standards.org.au/DevelopingStandards /AustralianStandardsGuidingPrinciples.aspx

U.S. Department of Energy. 2006. "Energy Conservation Standards Activities: Submitted Pursuant to Section 141 of the Energy Policy Act of 2005 and to the Conference Report (109-275) to the FY 2006

Energy and Water Development Appropriations Act." Available at: http://www1.eere.energy.gov/...standards/pdfs/implementationreport0806.pdf

U.S. Department of Energy. 2012. "Energy Conservation Program for Consumer Products: Test Procedures for Refrigerators, Refrigerator-Freezers and Freezers." Federal Register 77 FR 3559, January 25, 2012. Available at: http://www1.eere.energy.gov/buildings/appliance standards /pdfs/refr frz tp fr notice 01 25 12.pdf

U.S. Department of Energy Office of Energy Efficiency and Renewable Energy (EERE). 2009. "Fiscal Year 2009 Budget in Brief." Available at: http://www1.eere.energy.gov/ba/pba/pdfs/FY09 budget brief.pdf

Waide, P. 2010. "Opportunities for Success and CO2 Savings from Appliance Energy Efficiency Harmonization." London: Collaborative Labeling and Appliance Standards Program.

Waide, P. 2011. "Overview and Update of the ERP Directive, Energy Labelling Directive and Eco-label in the European Union." Presented at the Asian Energy Efficiency Standards and Labeling Forum. Guilin, China: 15 November 2011.

Wiel, S. and J.E. McMahon. 2005. "Energy Efficiency Labels and Standards: A Guidebook for Appliances, Equipment and Lighting (2nd edition)." Washington, DC: Collaborative Labeling and Appliance Standards Program (CLASP).

Wilkenfeld G. and Associates and Jacob M. Associates. 2010. "Consultation Regulation Impact Statement: National Legislation for Appliance and Equipment Minimum Energy Performance Standards (MEPS) and Energy Labeling." http://www.energyrating.gov.au/library/pubs/201001-consultation-ris-nationalMEPS-labelling.pdf

End Notes

[1] This comparison is simplified and do not necessarily reflect some key differences in test procedures for refrigerators, which may result in slightly different measured energy consumption for the same unit. More details on the differences in refrigerator test standards and comparability can be found in Fridley et. al. 2009.

[3] See "Energy Conservation Program: Certification, Compliance, and Enforcement for Consumer Products and Commercial and Industrial Equipment: Final Rule." Federal Register 76 (March 7, 2011). Available at: http://www1.eere.energy.gov/buildings/appliance standards/pdfs/cce finalrule notice.pdf

In: International Review ...
Editor: Chad R. Schmitz

ISBN: 978-1-62948-002-2

Chapter 2

DEVELOPMENT AND IMPLEMENTATION OF ENERGY EFFICIENCY STANDARDS AND LABELING PROGRAMS IN CHINA: PROGRESS AND CHALLENGES*

Nan Zhou, Nina Zheng Khanna, David Fridley and John Romankiewicz

EXECUTIVE SUMMARY

Over the last twenty years, with growing policy emphasis on improving energy efficiency and reducing environmental pollution and carbon emissions, China has implemented a series of new minimum energy performance standards (MEPS) and mandatory and voluntary energy labels to improve appliance energy efficiency. As China begins planning for the next phase of standards and labeling (S&L) program development under the 12th Five Year Plan, an evaluation of recent program developments and future directions is needed to identify gaps that still exist when compared with international best practices. The review of China's S&L program development and implementation in comparison with major findings from international experiences reveal that there are still areas of improvement, particularly when compared to success factors observed across leading international S&L program.

---

* This is an edited, reformatted and augmented version of Lawrence Berkeley National Laboratory, Publication No. LBNL-6141E, dated January 2013.

China currently lacks a formalized regulatory process for standard-setting and do not have any legal or regulatory guidance on elements of S&L development such as stakeholder participation or the issue of legal precedence between conflicting national, industrial and local standards. Consequently, China's laws regarding standard-setting and management of the mandatory energy label program could be updated, as they have not been amended or revised recently and no longer reflects the current situation.

While China uses similar principles for choosing target products as the U.S., Australia, EU and Japan, including high energy-consumption, mature industry and testing procedure and stakeholder support, recent MEPS revisions have generally aimed at only eliminating the bottom 20% efficiency of the market. Setting a firm principle based on maximizing energy savings that are technically feasible and economically justified may help improve the stringency of China's MEPS program and reduce the need for frequent revisions.

China also lacks robust survey data and relies primarily on market research data in relatively simple techno-economic analyses used to determine its efficiency standards levels rather than the specific sets of analyses and tools used internationally. Based on international experiences, inclusion of more detailed energy consumption surveys in the Chinese national census surveys and statistical reporting systems could help provide the necessary data for more comprehensive standard-setting analyses.

In terms of stakeholder participation in the standards development process, stakeholder participation in China is limited to membership on technical committees responsible for developing or revising standards and generally do not include environmental groups, consumer associations, utilities and other NGOs. Increasing stakeholder involvement to broader interest groups could help garner more support and feedback in the S&L implementation process.

China has emerged as a leader in a national verification testing scheme with complementary pilot check-testing projects, but it still faces challenges with insufficient funding, low local awareness amongst some regulatory agencies and resistance to check-testing by some manufacturers, limited product sampling scope, and testing inconsistency and incomparability of results. Thus, further financial and staff resources and capacity building will be needed to overcome these remaining challenges and to expand impacts evaluations to assess the actual effectiveness of implementation and enforcement.

1. INTRODUCTION

As a growing consumer of household appliances, lighting and electronic products, China has seen a steady rise in residential electricity use with 13% average annual growth since the 1980s. In 1989, China introduced MEPS for appliances but efficiency standards did not gain traction in the policy arena until the late 1990s when it was formally endorsed by legislation. Likewise, China did not establish a voluntary energy labeling program until 1999 and the mandatory information labeling program only came into existence in 2005.

Over the last twenty years, with growing policy emphasis on improving energy efficiency and reducing environmental pollution and carbon emissions, China has implemented a series of new minimum energy performance standards (MEPS) and mandatory and voluntary energy labels to improve appliance energy efficiency. As China begins planning for the next phase of standards and labeling (S&L) program development under the 12th Five Year Plan, an evaluation of recent program developments and future directions is needed to identify gaps that still exist when compared with international best practices.

2. REVIEW OF CHINA'S S&L PROGRAMS

2.1. Legal Framework

The concept of appliance efficiency standard was introduced in 1988 after the Standardization Law of China (中华人民共和国标准化法) was adopted during the Fifth Session of the Standing Committee of the Seventh National People's Congress. This law sets the basis for the formulation, implementation, management and other related responsibilities of efficiency standards and designates the State Council's administrative department of standardization to oversee an unified national mandatory efficiency standards program. As a result, the first batch of minimum energy performance standards (MEPS) was adopted in 1989 for eight major products, including refrigerators, room air conditioners, clothes washers, rice cookers and televisions. In 1995, the China National Institute of Standardization (CNIS) was authorized to organize MEPS development and revision. However, the standards were not formally endorsed in a legal manner until the 1997 National Energy Conservation Law of China

(中华人 民共和国节约能源法) provided the regulatory basis of mandatory energy efficiency standards for energy-consuming products and equipment and helped motivate program improvements. Specifically, Article 14 of the Law calls for the formulation of national standards by the administrative department for standardization under the State Council which manufacturers are legally bound to comply with under Article 24 (National People's Congress 2008). If manufacturers are found in violation of the standard, then the administrative department may submit a proposal to suspend production by the manufacturer and possibly mandate its closure under Article 43. The Energy Conservation Law also promoted the concept of energy efficiency labels in Article 18 and laid the regulatory foundation by discussing possible punitive measures for counterfeit labels under Article 48.

Subsequently in 1999, the China National Institute of Standardization (CNIS) began the process of revising single-period mandatory energy efficiency standards and developing new standards to follow international best practice while the China Standards Certification Center launched a new voluntary energy efficiency endorsement labeling program targeting the top 25% most efficient products. NDRC also issued the Management Method for Energy Conservation Products Certification in 1999 to establish the administrative framework for certifying standards and the voluntary endorsement label. In 2002, the General Administration of Quality Supervision, Inspection and Quarantine (ASQIQ) issued legal authority for its local offices to enforce safety, efficiency and other standards (APEC 2009).

The *Medium and Long-Term Plan for Energy Conservation* issued by NDRC in 2004 further supported the development and implementation of appliance standards and energy labels by addressing weaknesses with lagging efficiency standards, inadequate compliance and slack enforcement of Energy Conservation Law efforts (NDRC 2004). In particular, it encouraged more stringent revisions of energy efficiency standards, tightened market entry and the use of market mechanism to stimulate demand for energy-efficient consumer products as important principles of energy conservation. Household and office electric appliances were also included as a key subcategory for energy conservation within the building sector and implementation of energy efficiency standards and labeling was identified as a key project.

More recently, the mandatory categorical energy information label known as the China Energy Label was established in 2005 following legal provisions in the Energy Conservation Law with supporting regulation and support for implementation in the *Product Quality Law* and *Legislation on Certification &*

Accreditation (Jin and Li 2006). The administration of the China Energy Label program along with details on supervision and implementation, penalties and other supplementary provisions were established in the *Administration Regulation on Energy Efficiency Label* (Jin and Li 2006).

With the onset of China's carbon intensity target for 2020 and a ramp up in energy efficiency programs nationwide, the legal basis for China's standards and labeling programs has been strengthened in recent years. In the newly revised 2008 "Energy Conservation Law of People's Republic of China," 11 articles were related to MEPS and called for the examination and supervision of energy conservation regulations for high energy-consuming products. Article 13 also allowed local standards that are more stringent to be submitted to the State Council for approval. Under the revised "Energy Conservation Law" and the "Circular Economy Promotion Law of the People's Republic of China," provisions were also added relating to energy efficiency labeling, and the management mechanism and penalty methods have been clearly defined, thus establishing the legal foundation for incorporating the energy efficiency labeling system as an important national energy saving management measure. The legislative and regulatory history related to China's MEPS and labeling programs are illustrated in Figure 1.

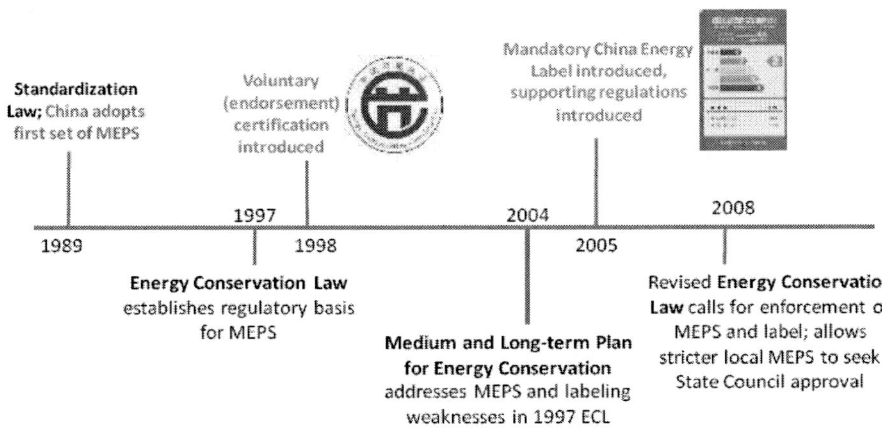

Figure 1. Legislative and Regulatory Timeline of China's MEPS and Labeling Programs, 1989 – Present.

Lastly, national and sectoral development plans also have important role in promoting mandatory energy efficiency standards. For example, China's 12th Five Year Plan for national economic and social development clearly indicates that "energy conservation and efficiency regulations and standards will be improved, formulation and implementation of product efficiency standards will be strengthened, and the China Energy Label, voluntary efficiency certification and compulsory government procurement programs will all be improved."

2.2. Standard Setting Process

In China, the principles for choosing target products for standard setting focus on the following characteristics:

- High energy consumption and high energy savings potential □ Widely used with mature industry and well-regulated market
- Mature testing procedure and good testing infrastructure and ability nationwide
- Stakeholder support

In terms of guiding principles for standard-setting, China has generally aimed at eliminating the bottom 20% efficiency of the market with each new standard and standard revision (Li 2012). More recently, economic and technical analysis are also used to help determine the proposed MEPS level. Unlike other countries such as the U.S. and European Union member states, China does not follow a specific legally prescribed standard-setting process. Rather, according to the 1990 Management Regulations for National Standards (国家标准管理办法), standards development and revision activities are organized and implemented by the national State Council administrative departments of standardization management and leadership and the Technical Committee. Specifically, they are responsible for supervising the progress of national standards development activities and ensuring completion of planned tasks, including the drafting of new or revised standards. The organization designated to draft the efficiency standards are responsible for providing the quality and technical content in accordance with GuoBiao (国标—GB) standard guidelines and also for preparing the accompanying Explanation of the Standard (编制说明). The Explanation of the Standard provides an overview and summary of the development of that particular standard and may include specific content on:

- Work summary: resources, designated organization, key stages of work, main organization for drafting standard
- Principles and main elements of the standard (e.g., technical indicators, parameters, formulae)
- Performance requirements, test procedures and implementation guidelines. For revised standards, a comparison of proposed revision and old standard levels is also provided.
- Key analyses, synthesis, technical and economic feasibility analyses and expected economic impacts
- Review of efficiency standards adopted elsewhere in the world and comparisons with existing international standards, or with foreign products and prototypes
- Relationship to existing laws, regulations and mandatory national standards
- Responses to major divergence in opinions and viewpoints
- Proposals to set national standards as mandatory or recommended standards
- Implementation of national standards and policy initiatives (including organizational measures, technical measures, transitional measures, etc.)
- Proposed repeal of existing standards;

After the draft standard, explanation of the standard and other related documents have been reviewed and finalized by the drafting organization, they are distributed to key organizations involved in the production, distribution, use, technical development and testing of the given product for comments and feedback. There is usually a set time period of approximately two months for comments, during which representatives should provide feedback, or state that they do not have significant comments. For relatively significant objections or arguments, the technical and economic feasibility and basis should also be provided. The standard drafting organization is then responsible for collecting, analyzing and synthesizing all comments in a summary of comments in the appendices of the explanation of the standard. This summary is then sent to the secretariat of the technical committee for review to determine if further review and a second period of commenting are needed.

Although there is not a set of legally prescribed procedures to follow in standards development in China, the main steps in a sample product standard-setting process in China are illustrated in Figure 6. From a technical perspective, once a target product has been identified and the standard development project approved by the Standardization Administration of China (SAC), the first analysis to be conducted is the market survey and market data collection. In the case of CFLs, questionnaire surveys were sent to 188 manufacturers with follow-up on-site visits to stores, lighting industry associations and other testing experts. Markets samples were also purchased and tested to determine the existing efficiency distribution of products in the market. Next, a techno-economic analysis is conducted to evaluate the potential energy savings and lifecycle costs of different efficiency design options as the basis for developing a draft standard. Although this analysis is known as "engineering/economic analysis" in China, it differs from the engineering analysis that is done in the U.S. standard-setting process in which efficient product models are deconstructed to evaluate the savings from each design element. In China, manufacturing costs and industrial impacts are estimated through interviews with manufacturers, rather than derived directly from modeling analysis. Lastly, the impact of proposed standard levels on national energy use and CO_2 emissions are estimated in the energy and environmental analysis using predicted sales and stock trends and unit energy consumption factors with and without the standard.

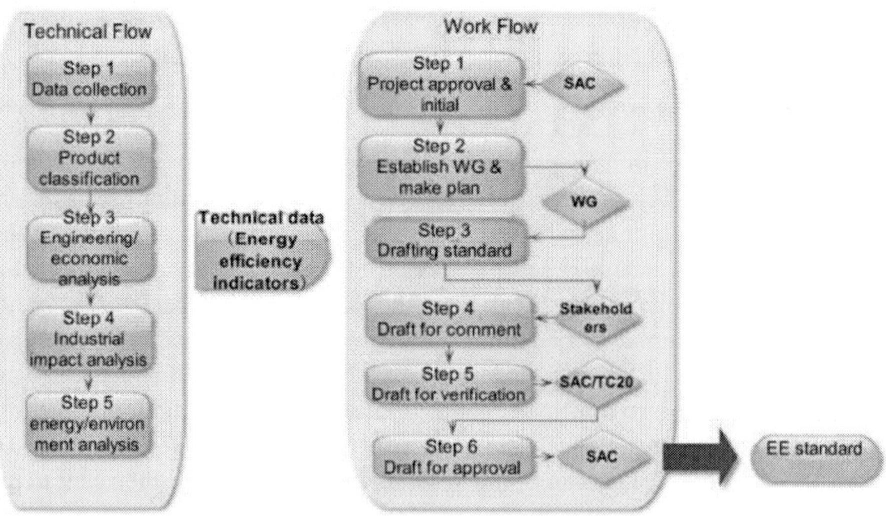

Source: Xia 2011.

Figure 2. China Efficiency Standard-setting Technical and Work Flow Process.

In terms of the regulatory steps for standard-setting, the draft standard is reviewed and comments by stakeholders are collected, reviewed and addressed during subsequent workshops. Once the standard has been approved, it is submitted to SAC as a national standard. For CFLs, the entire process occurred over 16 months, with 10 months for the initial technical analysis and draft standard development and 6 months for comments, verification and approval (Xia 2011). Typically, standard development and revisions process takes about 1.5 years, with the standard usually going into effect six months after the standard has been approved. This transitional period between the standard approval and implementation is determined by the SAC and may be longer or shorter than six months. Standards are then reviewed by either the technical committee or SAC within five years to determine if revisions are necessary to keep up with the pace of technical and efficiency improvements. If the standard does not need to be revised, then an amendment is added to state that the "standard was verified to remain effective in year XXXX." If the standard is revised, then the standard number remains the same but the revision year will be updated. If the standard is not longer needed, then the review unit will need to issue a written report detailing: review summary, main viewpoints of the reviewers and review findings. The standard can then be abolished after receiving the consent of the administrative departments overseeing national standards and the SAC.

Table 1. Summary Table of China's Standard Setting Process

Initial Legal Framework	National Standardization Law of 1989, National Energy Conservation Law of 1998
Regulatory Agency	State Administration of Quality, Supervision, Inspection, and Quarantine (AQSIQ), China National Institute of Standardization (CNIS)
Standard Setting Principle	Eliminate bottom 20% of market during each revision; use "reach" standards: two-period, two-tiered standards
Standard Setting Timeline	Standards are reviewed and revised as necessary usually within a period of five years, with the specific time of revision dependent on pace of technical and efficiency improvements. MEPS and mandatory China Energy Labels are also introduced for new products every year, but there is no pre-determined schedule for new or revised standards.
National versus Local Regulatory Precedence	National regulation has full mandate over MEPS and mandatory labeling

2.3. Analytical Tools and Methods

In the process of setting and revising mandatory energy efficiency standards, China generally utilizes techno-economic analysis and market impact analysis for evaluating efficiency levels. Based on international exchanges with comparable developed countries on their experiences, China's analytical methods and tools are considered to be consistent with tools used in the US standards setting process. But owing to the lack of data, financial support and comprehensive technological capabilities, China currently uses relatively simple analyses in determining its efficiency standards levels rather than the analytical models and tools used elsewhere. Results of the analysis are incorporated into the procedures for standards setting and revision and are submitted along with the draft standard to the Standardization Administration of the People's Republic of China (SAC) for review and approval. However, the preparatory documents for standard-setting are not publicly released.

Given that mandatory energy efficiency standards have had increasingly important impact on the economy and society in recent years, SAC may require that sufficient market and societal impact analysis and risk assessment be conducted as part of the standard setting and revision process in the near future. At the same time, SAC may also conduct public hearings on mandated standards for consumers, manufacturers, government authorities and other stakeholders to express their views. In recent years, the Chinese government has also increased investment in related fields of research and actively supported the research and development of related analytical models and tools.

2.4. Data Collection

At present, the Chinese National Bureau of Statistic does not survey end-use energy consumption and lacks residential and commercial energy consumption surveys) similar to those undertaken by the Energy Information Administration in the U.S (Residential Energy Consumption Survey or RECS and Commercial Building Energy Consumption Survey or CBECS). As a result, there is a lack of centralized data provided by an authoritative source on the energy use profiles of energy end-use products. At present, the market data, technology development trends, and data on the distribution of energy efficiency levels used in the current energy efficiency standard setting and revision processes are provided by market research agencies, are publicly available, are provided by manufacturers and companies, or are data from the

China Energy Label certification process. However, in the absence of a strong surveying and analysis framework and without mandatory legal provisions for manufacturers to report market data, it is difficult to guarantee the quality and reliability of the data currently used. At the same time, due to inadequate funding and financial support, the scope of the survey data used in the efficiency standards setting and revision process is also limited.

2.5. Stakeholder Participation

China's standard setting process currently involves limited stakeholder and public participation. The standard setting process is run by SAC, once technical recommendations have been made by CNIS. After an initial draft of the standard has been formulated by CNIS and approved by a working group convened by SAC, comments from stakeholders on the Technical Committees are collected during workshops and through written communication. For example, for the CFL lighting standard, two workshops were held in Jiangsu and Zhejiang with participants including members of the Working Group, research institutes and manufacturers. Nearly 40 comments were also collected over a two month period and each comment or suggestion was discussed in the Zhejiang workshop (Xia 2011). Other stakeholders that participate in the broader S&L process in China include industry associations, manufacturers, testing institutes, research institutes and universities while participation by environmental groups, consumer associations, utilities and other NGOs are relatively limited. Public meetings are not held currently but may be introduced in the future.

China currently has more than 470 Technical Committees, with the National Energy Foundation and Management of Standardization Technical Committee (SAC/TC20) responsible for energy efficiency standards. The Secretariat of SAC/TC20 is located in CNIS and the Committee is responsible for MEPS, rational use of energy standards, energy management standards, fuel economy standards, energy monitoring standards and other energy infrastructure and management standards. The Committee is composed of representatives from key government agencies, research institutes, industrial associations, leading universities and industrial enterprises.

2.6. Program Enforcement

In China, AQSIQ (State Administration of Quality, Supervision, Inspection and Quarantine) has the official authority over the supervision and the inspection of energy efficiency information, as established by the 1990 and 2005 revised Management Method for Energy Standardization. AQSIQ and its provincial divisions are responsible for national product quality supervision testing for all consumer products, with primary emphasis on product safety and secondary emphasis on product performance. In particular, AQSIQ conducts regular national product quality supervision testing every quarter as well as special product quality testing if there are product quality concerns. Because there are no specific standards or regulations on energy efficiency testing requirements, energy efficiency receives relatively low priority in national quality testing and the major appliances of clothes washers, refrigerators and air conditioners have only been tested one to three times from 2001 to 2006. For example, clothes washers were tested in the second quarter of 2005 but energy efficiency was not tested, and room air conditioners were tested in a special product quality test in early 2006 (Lin 2006).

In the regular national product quality testing, AQSIQ first establishes a prioritized list of products for testing with priority given to human health, personal and property safety, industrial products essential to the economy and products reported to have quality defects by consumers (Lin 2007). Product samples are then taken by inspection agencies and the relevant provincial departments. Manufacturers then confirm the selected product samples, the samples are tested and an inspection report is issued with the spot-testing results. If a product sample fails the product quality testing, it may be retested and mandatory rectification or punitive measures may be imposed if the sample still fails the retest (CNIS 2006). Previous interviews with Chinese experts and manufacturers have identified weaknesses in the national product quality testing that largely resulted from insufficient government funding, which was previously reported to be only half a million RMB per year for testing all appliances, home electronics and lighting products (Lin 2006). Small sample sizes, inadequate energy efficiency testing and infrequent testing of household appliances have all limited the role of national testing and monitoring of MEPS compliance.

Thus, “enterprise self-declaration” has historically been the key feature of MEPS and energy labeling enforcement as AQSIQ and the other organizations were not allocated sufficient resources to focus on energy efficiency testing. In recent years, however, several random market inspections and investigations of national and local supervision departments have raised questions about the validity of self-reported information as some enterprises and third-party laboratories were found to lack sufficient energy efficiency testing capacity (Zhou et al. 2010b). As a result, the central government has increased efforts to improve the enforcement and monitoring mechanisms for appliance energy efficiency standards in recent years with several energy efficiency testing and verification pilot programs undertaken by CNIS (see sections below).

Most recently in 2010, the China Energy Label Management Center within CNIS launched a special investigation to conduct check-testing of products from targeted manufacturers. Specifically, product samples from manufacturers with reported non-compliance and inconsistent results in the certification database and from the List of Leading Energy Efficiency Manufacturers were purchased from general circulation and sent to the National Quality Inspection Center for testing. The China Energy Label Management Center then suspended the label certification of non-compliant products and issued a deadline for rectification to the manufacturers. The compliance results from this special investigation in 2010 are shown in Table 2.

Table 2. 2010 Compliance Investigation of Selected Products

Product Type	Compliance Rate (%)	Non-Compliance Rate (%)	Total Samples (Units)
Storage Water Heaters	93%	7%	15
Alternate Current Fans	75%	25%	12
Gas Water Heater	100%	0%	10
Rice Cookers	93%	7%	15
Household Cooktops	80%	20%	15
Computer Display Monitors	85%	15%	20
Small and Medium-size Three Phase Asynchronous Motors	80%	20%	10
High Pressure Sodium Lamps	100%	0%	5

Source: CNIS 2012.

2.6.1. Pilot Energy Efficiency Check-testing Programs

In 2006, CNIS with funding from international organizations[1] launched an unprecedented sample energy efficiency checking and testing program to evaluate MEPS compliance and labeling accuracy. The 2006 program purchased and tested 43 models of refrigerators, room air conditioners and washers from retailers in the major cities of Guangzhou, Beijing and Hefei. The specific check-testing process is discussed further below. The first round of check-testing presented mixed results on the implementation and enforcement of the selected appliance standards, with a range of noncompliant product models in the three cities and overall products compliance rates between 71% and 91% (Zhou et al., 2008). The 2006 test results also showed notable variations in performance and noncompliance rates between product models sold in high-end, first-tier appliance retailers and second and third-tier retailers such as local appliance markets.

In 2007, appliance check-testing was repeated with a larger sample size of 73 models in the same cities. There was overall improvement in compliance rates for all three tested products with a significant drop in the number of non-compliant models from 11 out of 54 in 2006 to only 3 out of 73 models in 2007 (Zhou et al. 2008). On the regional level, compliance rate improvements were also realized in Beijing and unlike results from the previous year, no major variations in compliance were observed between different appliance markets in 2007. In August of 2009, CNIS launched the National and Local Enforcement of Energy Efficiency Standards and Labeling project with check-testing for a total of seven products conducted in four pilot locations of Jiangsu, Shandong and Sichuan provinces as well as Shanghai. This round of check-testing incorporated compliance verification for both MEPS and the China Energy Label. For verifying compliance with the China Energy Label, a set of inspection criteria was formulated and inspections were conducted at manufacturers' warehouses and selected retailers. During the inspections, results were recorded by product type and non-compliance was documented by photos of the product model and nameplate. For energy efficiency check-testing, the test procedure stated in each MEPS was followed with the overall check-testing process depicted in Figure 3.

Non-compliant manufacturers and retailers were subject to penalties based on the relevant laws and regulations and a list of non-compliant manufacturers were published on the China Energy Label website. In general, under the 2008 revised Energy Conservation Law, the specific possible penalties for violating mandatory labeling requirements are detailed and include fines of 10,000 to 30,000 RMB if the label is absent or do not comply with the specific labeling requirements and fines of 50,000 to 100,000 RMB for misleading or falsifying labeling results (NPC 2008). For serious violations, the business license of the manufacturer may also be revoked. For MEPS violations, enterprises that manufacturer, import or sell non-compliant products are ordered to stop production, import or sales and illegal gains from noncompliant products will be confiscated and the violator may be further fined 100% to 500% of the illegal proceeds from non-compliant products (NPC 2008).

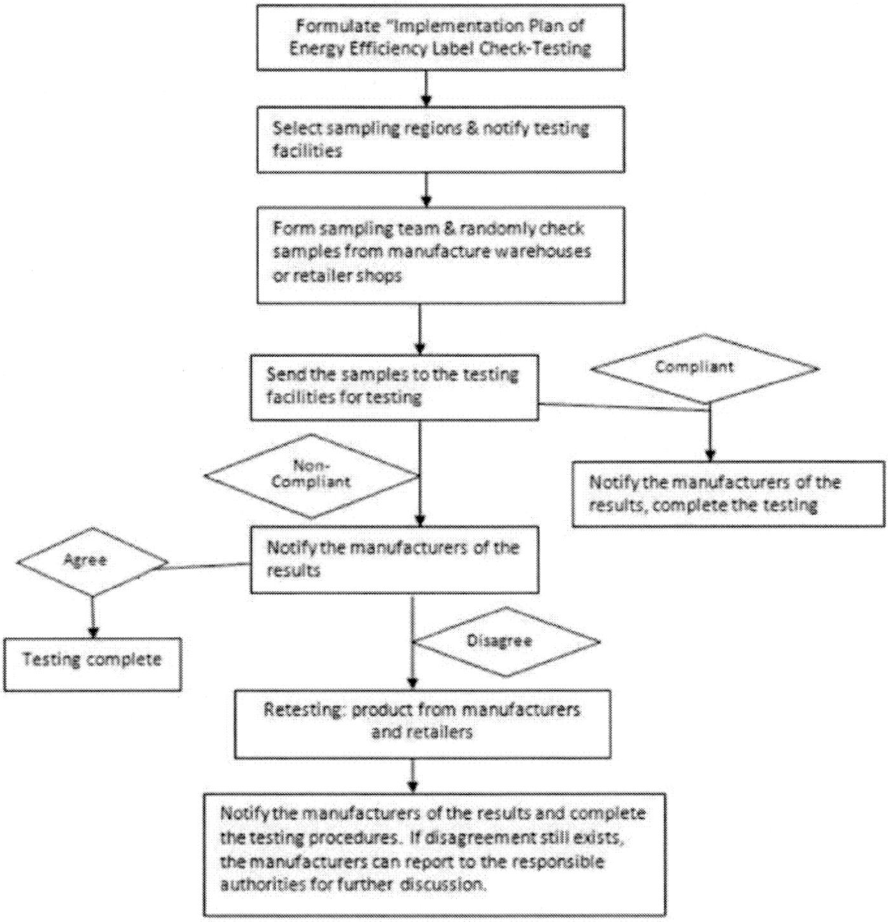

Source: Zhou et al. 2011.

Figure 3. Energy Efficiency Sampling and Testing Process for China's Pilot Check-testing Programs.

The 2009 round of testing found that Sichuan province had the lowest overall compliance for both standards and labeling. Although the specific reasons for lower compliance could not be directly identified, possible challenges with enforcement in Sichuan could include slower economic development and lower awareness, staff expertise and funding to support standards and labeling enforcement (Zhou et al. 2011). Other challenges identified in the 2009 testing include lack of awareness and lack of an initial publicity campaign which resulted in manufactures' failing to register their products; retailers' resistance to inspection; lack of timely updating of product information online; incidents of different laboratories reaching different results for the same product; higher compliance rates for products produced by large manufacturers along with lack of attention to enforcement for smaller manufacturers (Zhou et. al. 2011).

2.6.2. Pilot Round-Robin Testing Program

The early check-testing rounds highlighted sample selection concerns with uneven distribution in the grade of testing samples selected for each product and inconsistent test results with significant variations in results when tested in different laboratories. In order to identify the possible differences between laboratories and to improve consistency, reliability, accuracy, and credibility of the China Energy Label, an inter-laboratory round robin testing (RRT) program was launched by CNIS in 2009. A leading domestic manufacturer was asked to produce three sets of split air conditioners, with an additional sample initially tested in Australia, and the samples were sent to six Chinese laboratories and a Japanese laboratory for efficiency testing following the MEPS. In the end, 43 tests were completed in 4 Chinese laboratories and the results showed a good level of quality for the measurement of the efficiencies and the capacities of air/air air conditioners (Saheb et al. 2010). Nevertheless, the test also revealed room for improvement in specific testing methods within specific laboratories and the transportation, installation and setting of samples.

Table 3. Summary of China's National Product Quality Testing and Pilot Energy Efficiency Check-testing Programs

	China
Certification requirements	Self-reported manufacturer certification for China Energy Label
Check-testing	May be included in national product quality testing organized by AQSIQ. *Pilot energy efficiency check-testing program focused on MEPS compliance launched in selected provinces in 2006, 2007 and 2009.*
Sample Selection Method	Samples randomly selected from manufacturers' warehouses and/or retailers.

	China
Sample Size	Varies by product and type of quality testing (regular versus special). *Pilot energy efficiency check-testing varies by province and product and ranged from 5 to 30 batches of samples in 2009.*
Testing Process	Two stages of testing, with re-testing in Stage 2 requested by manufacturers. *For 2009 pilot energy efficiency check-testing, testing done on batches of samples.*
Compliance verification	National product quality testing primarily focuses on product safety and performance, with minimal emphasis on energy efficiency. *Pilot energy efficiency check-testing focused on energy efficiency testing and MEPS compliance, but 2009 pilot check-testing also verified labeling compliance.*
Fines/penalty for Non-compliance	Negligent or false use of voluntary label has RMB 30,000-100,000 penalty; no label or false use of mandatory label has a fine of RMB 10,000-30,000 and 50,000-100,000, respectively; For MEPS, companies can be fined 1-5 times any illegal gains made from products sold that are not in compliance with MEPS.
Testing budget	Historically, AQSIQ has been allocated 500,000 RMB per year for national product quality testing of all household appliances, home electronics and lighting products. *In pilot energy efficiency check-testing, 200,000 RMB allocated to each pilot region with additional regional funding expected.*
Informal enforcement mechanisms	None
Information sharing between agencies or jurisdictions	None
Voluntary certification programs	Voluntary energy efficiency certification program managed by China Quality Certification Centre.

Note: Italics indicate enforcement elements that only apply to pilot energy efficiency check-testing programs.

The pilot energy efficiency check-testing and round-robin testing are particularly significant given that China remains a rapidly developing economy, yet developed economies still lack robust and regular enforcement testing. For example, the U.S. does not have a national check-testing enforcement program for MEPS and only recently launched enforcement testing program for ENERGY STAR along with round-robin testing. At the same time, however, expanding China's verification testing programs to cover more models and products and developing a plan for ramping up the national verification testing program over the next three to five years are important next steps in strengthening appliance standards and labeling implementation. Further improvement in enforcement could also come about through more awareness and educational campaigns both at the national and provincial level; more publicity for those who continuously excel in compliance and those who fail; the designation of specific funding and staffing from central and provincial governments for local enforcement projects; emphasis on targeting non-compliant manufacturers in subsequent years; and greater emphasis on implementing standardized testing tools and procedures among different testing laboratories. This will be particularly important as the information labeling program gains more visibility and expands to additional product categories.

2.7. Basis for Test Procedures

The test procedures for products are typically developed and set by the product-specific or industry-specific standardization technical committees and the test procedures will be referenced and integrated into the relevant sections of the energy efficiency standard during the standard setting process. Since a guiding principle of standards development in China is encouraging the adoption and revision of ISO or IEC international standards on the basis of domestic conditions, the product standardization technical committees will generally adopt or revise relevant ISO or IEC standards during the test procedure development process after evaluating if it is more suitable and appropriate to adopt existing international standards. At the same time, the technical committees may also elect to develop a China-specific test procedure that differs from existing international test procedures if it is believed to better represent national conditions or criteria, as was the case for televisions. In the absence of existing international test procedures - such as the case for products not yet covered by international standards and labeling programs - the technical committees will develop a new test procedure for China. Due to the differing product timeline for standard development, if a standardization technical committee has not initiated test procedure development process when efficiency standards setting is undertaken, then related ISO or IEC standards are revised and used in China. In recent years, due to the fast pace development of technology, China have referenced test procedures from EnergyStar, European Union's EcoLabel and other international product associations such as the International Commission on Illumination (CIE) in the standard setting process.

2.8. Program Resources

In China, AQSIQ is allocated RMB 500,000 per year for the standards development and testing of all household appliances, home electronics, and lighting products. Most of this funding was used to pay for the national product quality supervision testing fees, in the range of RMB 200,000-300,000. Of the total AQSIQ funding, CNIS and its Standardization Committee receive about RMB 20,000 to 30,000 for the development of one standard.

In addition, CNIS has historically received 50,000 to 100,000 USD from international organizations such as the Energy Foundation for the development of one standard. Because AQSIQ is responsible for all types of product standards and treat each type equally and NDRC does not have the authority to directly increase the funding for energy efficiency standards development, national funding for MEPS development and enforcement is relatively limited.

However, it is possible for NDRC and the central government to provide more funding to the China Energy Labeling program as it currently receives no direct government funding. Rather, the China Energy Label program is supported solely by international funding such as the Energy Foundation, and the United Nations End-Use Energy Efficiency Programme (EUEEP) and Global Environmental Fund program. While China has significantly increased the government funding for energy efficiency with growth in funding from RMB 9 billion in 2007 to RMB 36 billion in 2009 (Zhou et al. 2010a), most of the funding related to appliance energy efficiency was used for subsidy and rebate programs such as "Appliances to the Countryside (家电下乡)" and the "Energy-Efficient Product Subsidy Program (惠民I程). This suggests that the financial resources provided by the central government for appliance standards and labeling programs are insufficient and the programs could be strengthened with additional government funding.

3. Summary of Key Findings from International Review

A separate LBNL report provides in-depth review and comparative analysis of the development and recent advancements in the U.S. MEPS and Energy Star, Australia MEPS and Energy Label, European Union MEPS and Eco-Design Label and Japanese Top Runner programs. Specific details on the individual programs and the comprehensive comparison of the four regions can be found in Zhou et al. 2012. Based on this comparative analysis, best practice examples of key programmatic elements as well as cross-cutting factors for success and lessons learned in S&L program development and implementation are identified. In particular, the comparative review of standards and labeling programs in the U.S., Australia, EU and Japan has uncovered some overarching themes and highlighted several key factors to successful program elements:

Sound Legal Framework for Standard-Setting

Standard-setting and programmatic implementation can benefit significantly from a legal framework that directly specifies a timeline or schedule for standard-setting and revision, product coverage and legal sanctions for non-compliance.

Standard-setting Principle Focused on Maximizing Feasible Energy Savings

Programs in the four regions revealed similarities in guiding principles of standards development that focus on achieving significant energy savings that are technically feasible and economically justified. This helps ensure that future MEPS will be set at levels that exceed the average product efficiency on the market and will effectively push the market forward towards higher efficiency levels once implemented.

Data and Analytical Support for Comprehensive Standard-Setting Analysis

In terms of analytical support for standard-setting, detailed survey data such as the U.S. Residential Energy Consumption Survey and rigorous sets of technical, economic, energy and environmental analyses provide a strong foundation for setting a particular standard level.

Active Stakeholder Participation in Standards Development and Implementation Process

Stakeholder participation also strengthens the standard-setting process by incorporating insights from different groups of stakeholders, although the particular form of participation may vary from product-specific stakeholder forums to membership on technical committees to public comments to technical analyses.

Effective Enforcement and Program Resources

Sufficient program resources are critical to the effectiveness of standards and labeling programs and cost-sharing between national and local governments can be undertaken to ensure adequate resources and uniform implementation. Resources are needed to support enforcement mechanisms such as check-testing and monitoring, which in turn impact the effectiveness of punitive measures such as cancellation of registration or product sales-based fines as effective deterrents for non-compliance.

In addition to these overarching factors for success, the international review of the four selected regions also illustrates that while no single country has best practices in all elements of standards and labeling development and implementation, national examples of best practices for individual elements do exist. For example, the U.S. has demonstrated rigorous analyses for standard-setting and robust data source with the RECS database while Japan's Top Runner standard-setting principle has been effective in motivating manufacturers to exceed targets ahead of time. In terms of standards implementation and enforcement, Australia has demonstrated success in enforcement with its long history of check-testing and enforcement initiatives while mandatory information-sharing between EU jurisdictions on compliance results is another important enforcement mechanism. As reflected by these examples, it is important to understand not only the drivers of different paths of standards and labeling development, but also the country-specific context for examples of best practices in understanding why certain S&L programs have been effective.

4. Implications for China from International Review

4.1. Legal Framework

As an important legal foundation for appliance efficiency standards in China, the *Standardization Law of China* has not been amended or revised in a long time and thus lags behind market and technological developments since its passage. Likewise, the China Energy Label Management Law promulgated along with the 2007 revision of the Energy Conservation Law could also be updated to reflect rapid changes in the labeling program over the past few years. This is especially relevant given the lack of specific legal or regulatory requirements for China's entire standard-setting and revision process, stakeholder involvement and participation, and implementation and enforcement of the mandatory MEPS and labeling programs. In addition, the coordination and legal precedence between national standards, industrial standards and local standards is another area that has yet to be addressed. Consequently, China's laws regarding standard-setting and management of the mandatory energy label program may require updating, as they have not been amended or revised recently and no longer reflects recent developments and the current situation.

4.2. Standard-setting Principles

China uses similar principles for choosing target products, including high energy-consumption, mature industry and in-place testing procedures and stakeholder support, but recent MEPS revisions have generally aimed at only eliminating the bottom 20% efficiency of the market. Setting a firm principle based on maximizing energy savings that are technically feasible and economically justified may help improve the stringency of China's MEPS program and reduce the need for frequent revisions.

4.3. Standard Development Analyses and Data

Technical analyses and data availability provide underlying foundation for the development of effective efficiency standards and labeling programs. Currently, China lacks capacity in both data collection and a comprehensive framework for standard-setting analysis. Because there is no designated and authoritative source for necessary data inputs to technical analyses, including basic data on end-use usage patterns and energy consumption patterns and detailed sales data by efficiency level, standard-setting analyses are largely dependent on simplified market data that may not always be consistent. Based on international experiences, inclusion of more detailed energy consumption surveys in the Chinese national census surveys and statistical reporting systems could help provide the necessary data for more comprehensive standard-setting analyses.

Due to this lack of data as well as inadequate financial support and technological capabilities, China currently uses relatively simple techno-economic analyses in determining its efficiency standards levels rather than the specific sets of analyses and tolls used internationally. For example, compared with other major analyses undertaken in international standards development, China lacks capacity and experience in engineering analysis, energy and water use analysis, mark-up analysis, shipment analysis, national impact analysis, manufacturer impact analysis, life-cycle cost subgroup analysis, employment impact analysis, regulatory impact analysis and environmental assessment analysis. While not all of these analyses may be necessary in standard-setting in China, a more comprehensive analytical approach encompassing at least some of these analyses need to be developed and implemented on a consistent basis. More research is needed to identify the most essential and appropriate analyses that should be included in a standard-setting framework for China.

4.4. Stakeholder Participation

Compared to the international experiences of the four selected regions, stakeholder participation in China is limited to membership on technical committees responsible for developing or revising standards, which generally do not include environmental groups, consumer associations, utilities and other NGOs. Expanding participation to more stakeholders in the standards development and implementation process will help ensure there is not only broad awareness and understanding of China's S&L programs amongst local regulatory agencies, manufacturers and industrial partners, retailers, and consumers, but also continuous support and feedback to the programs.

4.5. Government Resource and Support

As an important government policy tool for improving energy efficiency, energy efficiency standards and labeling programs need the government's strong support in many aspects. At present, the management, guidance and support for the relevant government agencies responsible for implementing the energy efficiency standards and labeling programs in China are insufficient. Energy efficiency standards inherently have economic externalities given its benefits the public, and therefore often receive financial resources from governments in developed economies to support standards and labeling development and implementation. In China, however, the unstable and insufficient financial resources for S&L programs has resulted in delays in the necessary revisions of existing standards and the lack of a strong technical and scientific basis for the development of some new standards. At the same time, insufficient financial resources have also limited public awareness of standards and evaluations of the impact of these programs.

4.6. Program Enforcement and Resources

While China has emerged as a leader in a national verification testing scheme with complementary pilot check-testing projects, it still faces challenges in enforcement. On the national level, the main regulatory agency responsible for verification testing has limited resources and too many conflicting priorities, resulting in little emphasis on energy efficiency. The smaller scale pilot check-testing projects launched by the China National Institute of Standardization responsible for developing standards and managing the mandatory energy label face insufficient funding, low local awareness amongst some regulatory agencies and resistance to check-testing by some manufacturers, limited product sampling scope, and testing inconsistency and incomparability of results. Further financial and staff resources and capacity building will be needed to overcome these remaining challenges to enforcement and to expand impacts evaluations to assess the actual effectiveness of implementation and enforcement.

ACKNOWLEDGMENTS

This work was supported by the China Sustainable Energy Program of the Energy Foundation and the Collaborative Labeling and Appliance Standards Program through the U.S. Department of Energy. The authors are also very grateful to the China National Institute of Standardization and particularly Pengcheng Li for providing valuable insight related to this work.

REFERENCES

Asia-Pacific Economic Cooperation (APEC). 2009. "People's Republic of China Legislative History Related to Standards & Labels." http://www.apec-esis.org/countryoverview.php?country=People%27s%20 Republic%20of%20China&ID

China National Institute of Standardization. 2006. "The Research on China's Energy Efficiency Standards Implementation and Monitoring System." China Sustainable Energy Program Project Report. Available at: http://www.efchina.org/FReports.do?act=detail&id=228

Jin, M. and A. Li. 2006. *The Implications and Impacts of China Energy Label.* Proceedings of the 4th International Conference on Energy Efficiency in Domestic Appliances and Lighting, London, 21 – 23 June 2006.

Li Pengcheng. 2012. Personal Communication. 11 December 2012.

Lin, J. 2007. "Creating and Implementing a Regularized Monitoring and Enforcement System for China's Mandatory Standards and Energy Information Label for Appliances." LBNL Report. Berkeley, CA: Lawrence Berkeley National Laboratory.

National Development and Reform Commission (NDRC), 2004. *Medium and Long-Term Plan for Energy Conservation*. Nov 25, 2004.

National People's Congress, 1997. *Energy Conservation Law of China.* Approved by the 28th Standing Committee Meeting of the 8th National People's Congress. 1 November 1997.

National People's Congress. 2008. "Energy Conservation Law of the People's Republic of China (Revised)." Beijing: National People's Congress. 1 April 2008.

Saheb, Y., Zhou, N., Fridley, D. and A. Pierrot. 2010. "Compliance and Verification of Standards and Labeling Programs in China: Lessons Learned." LBNL Report 4599-E. Berkeley, CA: Lawrence Berkeley National Laboratory.

Xia, Y. 2011. "Case Study: Development and Implementation of CFL ES&L in China." Presented at 2011 2nd BRESL Training Workshop. Guilin, China: 14 November 2011.

Zhou, N., Zheng, N., Fridley, D., Wang, R. and C. Egan. 2008. "Check-testing of Manufacturer Self-Reported Labeling Data & Compliance with MEPS. LBNL Report 247-E. Berkeley, CA: Lawrence Berkeley National Laboratory.

Zhou, N., Levine, M. and L. Price. 2010a. Overview of Current Energy Efficiency Policies in China. Energy Policy 38 (11): 6439-6452.

Zhou, N., Fridley, D. and N. Zheng. 2010b. China Energy Efficiency Round Robin Testing Results for Room Air Conditioners. LBNL Report 3502-E. Berkeley, CA: Lawrence Berkeley National Laboratory.

Zhou, N., Zheng, N., Fino-Chen, C. and D. Fridley. 2011. "Status of the Local Enforcement of Energy Efficiency Standards and Labeling Program in China." LBNL Report 5289-E. Berkeley, CA: Lawrence Berkeley National Laboratory.

Zhou, N., Zheng, N., Fridley, D. 2012. "International Review of the Development and Implementation of Energy Efficiency Standards and Labeling Programs." LBNL Report 5407-E. Berkeley, CA: Lawrence Berkeley National Laboratory.

End Note

[1] China does not have a specified budget for MEPS enforcement and testing as the responsibilities often fall to local jurisdictions with varying financial capacities. The CNIS verification testing and round-robin testing efforts described were primarily funded through research grants from international organizations and foundations including the Japanese Ministry of Economics, Technology and Industry, the Energy Foundation's China Sustainable Energy Program, ClimateWorks Foundation and the Collaborative Labeling and Appliance Standards Program.

INDEX

D

E

F

G

H

W